LONDON MATHEMATICAL SOCIETY LECTURE NOTE SERIES

Managing Editor: Professor J.W.S. Cassels, Department of Pure Mathematics and Mathematical Statistics, University of Cambridge, 16 Mill Lane, Cambridge CB2 1SB, England

The books in the series listed below are available from booksellers, or, in case of difficulty, from Cambridge University Press.

4 Algebraic topology, J.F. ADAMS
5 Commutative algebra, J.T. KNIGHT
16 Topics in finite groups, T.M. GAGEN
17 Differential germs and catastrophes, Th. BROCKER & L. LANDER
18 A geometric approach to homology theory, S. BUONCRISTIANO, C.P. ROURKE & B.J. SANDERSON
20 Sheaf theory, B.R. TENNISON
21 Automatic continuity of linear operators, A.M. SINCLAIR
23 Parallelisms of complete designs, P.J. CAMERON
24 The topology of Stiefel manifolds, I.M. JAMES
25 Lie groups and compact groups, J.F. PRICE
27 Skew field constructions, P.M. COHN
29 Pontryagin duality and the structure of LCA groups, S.A. MORRIS
30 Interaction models, N.L. BIGGS
31 Continuous crossed products and type III von Neumann algebras, A. VAN DAELE
34 Representation theory of Lie groups, M.F. ATIYAH *et al*
36 Homological group theory, C.T.C. WALL (ed)
38 Surveys in combinatorics, B. BOLLOBAS (ed)
39 Affine sets and affine groups, D.G. NORTHCOTT
40 Introduction to $H_p$ spaces, P.J. KOOSIS
41 Theory and applications of Hopf bifurcation, B.D. HASSARD, N.D. KAZARINOFF & Y-H. WAN
42 Topics in the theory of group presentations, D.L. JOHNSON
43 Graphs, codes and designs, P.J. CAMERON & J.H. VAN LINT
44 Z/2-homotopy theory, M.C. CRABB
45 Recursion theory: its generalisations and applications, F.R. DRAKE & S.S. WAINER (eds)
46 p-adic analysis: a short course on recent work, N. KOBLITZ
48 Low-dimensional topology, R. BROWN & T.L. THICKSTUN (eds)
49 Finite geometries and designs, P. CAMERON, J.W.P. HIRSCHFELD & D.R. HUGHES (eds)
50 Commutator calculus and groups of homotopy classes, H.J. BAUES
51 Synthetic differential geometry, A. KOCK
52 Combinatorics, H.N.V. TEMPERLEY (ed)
54 Markov processes and related problems of analysis, E.B. DYNKIN
55 Ordered permutation groups, A.M.W. GLASS
56 Journees arithmetiques, J.V. ARMITAGE (ed)
57 Techniques of geometric topology, R.A. FENN
58 Singularities of smooth functions and maps, J.A. MARTINET
59 Applicable differential geometry, M. CRAMPIN & F.A.E. PIRANI
60 Integrable systems, S.P. NOVIKOV *et al*
61 The core model, A. DODD
62 Economics for mathematicians, J.W.S. CASSELS
63 Continuous semigroups in Banach algebras, A.M. SINCLAIR
64 Basic concepts of enriched category theory, G.M. KELLY
65 Several complex variables and complex manifolds I, M.J. FIELD
66 Several complex variables and complex manifolds II, M.J. FIELD
67 Classification problems in ergodic theory, W. PARRY & S. TUNCEL
68 Complex algebraic surfaces, A. BEAUVILLE
69 Representation theory, I.M. GELFAND *et al*

70 Stochastic differential equations on manifolds, K.D. ELWORTHY
71 Groups - St Andrews 1981, C.M. CAMPBELL & E.F. ROBERTSON (eds)
72 Commutative algebra: Durham 1981, R.Y. SHARP (ed)
73 Riemann surfaces: a view towards several complex variables, A.T. HUCKLEBERRY
74 Symmetric designs: an algebraic approach, E.S. LANDER
75 New geometric splittings of classical knots, L. SIEBENMANN & F. BONAHON
76 Spectral theory of linear differential operators and comparison algebras, H.O. CORDES
77 Isolated singular points on complete intersections, E.J.N. LOOIJENGA
78 A primer on Riemann surfaces, A.F. BEARDON
79 Probability, statistics and analysis, J.F.C. KINGMAN & G.E.H. REUTER (eds)
80 Introduction to the representation theory of compact and locally compact groups, A. ROBERT
81 Skew fields, P.K. DRAXL
82 Surveys in combinatorics, E.K. LLOYD (ed)
83 Homogeneous structures on Riemannian manifolds, F. TRICERRI & L. VANHECKE
84 Finite group algebras and their modules, P. LANDROCK
85 Solitons, P.G. DRAZIN
86 Topological topics, I.M. JAMES (ed)
87 Surveys in set theory, A.R.D. MATHIAS (ed)
88 FPF ring theory, C. FAITH & S. PAGE
89 An F-space sampler, N.J. KALTON, N.T. PECK & J.W. ROBERTS
90 Polytopes and symmetry, S.A. ROBERTSON
91 Classgroups of group rings, M.J. TAYLOR
92 Representation of rings over skew fields, A.H. SCHOFIELD
93 Aspects of topology, I.M. JAMES & E.H. KRONHEIMER (eds)
94 Representations of general linear groups, G.D. JAMES
95 Low-dimensional topology 1982, R.A. FENN (ed)
96 Diophantine equations over function fields, R.C. MASON
97 Varieties of constructive mathematics, D.S BRIDGES & F. RICHMAN
98 Localization in Noetherian rings, A.V. JATEGAONKAR
99 Methods of differential geometry in algebraic topology, M. KAROUBI & C. LERUSTE
100 Stopping time techniques for analysts and probabilists, L. EGGHE
101 Groups and geometry, ROGER C. LYNDON
102 Topology of the automorphism group of a free group, S.M. GERSTEN
103 Surveys in combinatorics 1985, I. ANDERSON (ed)
104 Elliptic structures on 3-manifolds, C.B. THOMAS
105 A local spectral theory for closed operators, I. ERDELYI & WANG SHENGWANG
106 Syzygies, E.G. EVANS & P. GRIFFITH
107 Compactification of Siegel moduli schemes, C-L. CHAI
108 Some topics in graph theory, H.P. YAP
109 Diophantine Analysis, J. LOXTON & A. VAN DER POORTEN (eds)
110 An introduction to surreal numbers, H. GONSHOR
111 Analytical and geometric aspects of hyperbolic space, D.B.A.EPSTEIN (ed)
112 Low-dimensional topology and Kleinian groups, D.B.A. EPSTEIN (ed)
113 Lectures on the asymptotic theory of ideals, D. REES
114 Lectures on Bochner-Riesz means, K.M. DAVIS & Y-C. CHANG
115 An introduction to independence for analysts, H.G. DALES & W.H. WOODIN
116 Representations of algebras, P.J. WEBB (ed)
117 Homotopy theory, E. REES & J.D.S. JONES (eds)
118 Skew linear groups, M. SHIRVANI & B. WEHRFRITZ
119 Triangulated categories in the representation theory of finite-dimensional algebras, D. HAPPEL
120 Lectures on Fermat varieties, T. SHIODA
121 Proceedings of *Groups - St Andrews 1985*, E. ROBERTSON & C. CAMPBELL (eds)
122 Non-classical continuum mechanics, R.J. KNOPS & A.A. LACEY (eds)
123 Surveys in combinatorics 1987, C. WHITEHEAD (ed)
124 Lie groupoids and Lie algebroids in differential geometry, K. MACKENZIE
125 Commutator theory for congruence modular varieties, R. FREESE & R. MCKENZIE

London Mathematical Society Lecture Note Series. 123

# Surveys in Combinatorics 1987

## Invited papers for the Eleventh British Combinatorial Conference

Edited by C. WHITEHEAD
Department of Mathematical Sciences, Goldsmiths' College, London

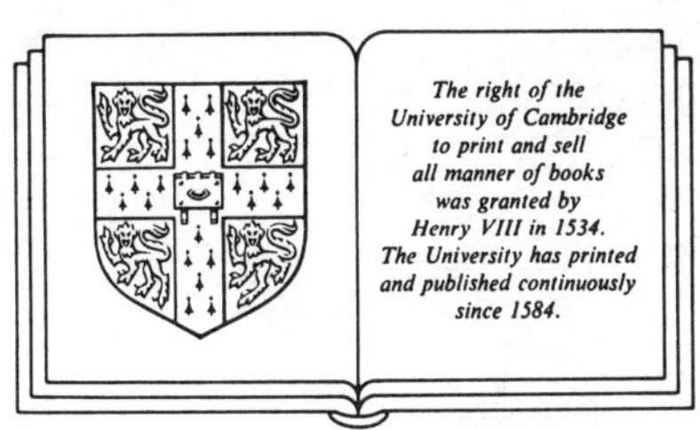


CAMBRIDGE UNIVERSITY PRESS
Cambridge
New York New Rochelle Melbourne Sydney

Published by the Press Syndicate of the University of Cambridge
The Pitt Building, Trumpington Street, Cambridge CB2 1RP
32 East 57th Street, New York, NY 10022, USA
10, Stamford Road, Oakleigh, Melbourne 3166, Australia



First published 1987

Printed in Great Britain at the University Press, Cambridge

*Library of Congress cataloging in publication data available*

*British Library cataloguing in publication data*
British Combinatorial Conference *(11th: 1987: Goldsmiths' College)*
Surveys in combinatorics, 1987: Invited papers for the Eleventh
British Combinatorial Conference. — (London Mathematical
Society lecture note series, ISSN 0076-0552; 123).
1. Combinatorial analysis
I. Title II. Whitehead, C. III. Series
511'.6 QA164

ISBN 0 521 34805 6

# CONTENTS

# PREFACE

The British Combinatorial Conferences are held biennially at different universities in the U.K. and regularly attract participants from all over the world. This volume contains the papers of the nine invited speakers at the Eleventh Conference, held from 13-17 July 1987, at the University of London Goldsmiths' College.

A tradition has recently been established at these Conferences to name one lecture after **Richard Rado** to commemorate his contribution to British combinatorics. **P. Erdős** was invited to give the Rado lecture at this Conference. His paper on his joint work with Richard Rado is a tribute to a friendship and collaboration spanning half a century. It gives the author's own view of the relative importance of their results not only to combinatorics but also to other related fields, such as set theory, topology and logic. A number of interesting problems that they have posed are still not completely solved and in this paper, Erdös gives the latest price for the scalp of each.

The eight other papers cover a number of areas of current research interest in combinatorics. **A. Barlotti** describes how methods of a purely geometric nature can be used to construct *designs*. He collects together a variety of these methods from a wide literature. **P.J. Cameron** reports on some very recent investigations into *sum-free sets*. He considers what can be said about the density of a sum-free set, mentioning a number of areas from which a "typical" sum-free set may arise and the directions that future investigations might take. **V. Chvátal** records progress in the problem of trying to characterize *perfect graphs*, an area of research initiated by a conjecture of C. Berge (1960). **P. Frankl** gives a detailed account of the use of the *shifting technique* in extremal set theory. **V. Rödl** and his co-author **R.L. Graham** describe the current state of knowledge in *Ramsey theory* and the methods needed to establish the bounds they discuss. **E.C. Milner**, in a joint paper with **K. Prikry**, describes recent work on *almost disjoint sets* which will be of interest to research workers in point-set topology as well as in combinatorics and set theory. The last two papers in this volume are on different aspects of graph theory. **A. Thomason** writes on *pseudo-random* graphs, an umbrella term used to describe random and strongly regular graphs on those occasions where they bear similarities. P. Winkler examines the *path metric* on the vertices of a graph and its simplification by means of isometric embedding in cartesian products of graphs.

I am grateful to a number of people who have helped me in editing this volume. My thanks go to the authors for producing their papers within the necessary timetable and the referees for their helpful comments and for noting a number of small misprints which might otherwise have gone undetected. I am particularly indebted to Ronald Graham of A T & T Bell Laboratories for arranging for the typing on their word processor of two papers in addition to the one of which he is a joint author and for his greatly valued help in editing these papers. But for his time and effort, so generously given, this book would be the poorer. Thanks are also due to Zoltan Füredi of M.I.T. for his help in typing and editing one of the papers and to Karen Young of the Mathematical Sciences Department, Goldsmiths' College for her assistance in word processing this Preface and the Contents list. I am grateful, too, to Martin Gilchrist of the Science Editorial staff of Cambridege University Press for his unfailing patience and encouragement.

The contributed papers at the Conference will again be published in a special issue of Ars Combinatoria.

London
30 March, 1987

C.A.W.

# FINITE GEOMETRIES AND DESIGNS

A.Barlotti
Istituto Matematico "U.Dini", Università di Firenze, Italy

Abstract. The purpose of this little expository paper is to present some of the connections between the theory of finite geometries (i.e. planes and spaces of various kinds) and design theory, stressing in particular the help that geometries offer in the construction of designs.

1. The modern Design Theory originated from the work done in statistics by R.A.Fisher and his associates. For the early history of the Design of Experiments cf., e.g., Bose 1947. In the spirit of Dembowski 1968 we consider the various types of designs as geometric structures. So we shall use Dembowski's terminology (for the translation of other terms frequently used see the "Dictionary" in Dembowski's book). We assume that the basic definitions, and the general properties, of designs are known. If not these can be easily found in one of the various books quoted in the references.

Our main purpose is to survey methods of a purely geometric nature that can be used for the construction of designs (see also Barlotti 1985). Our hope is that our description will lead some of the readers to find new nice and useful constructions of geometric type.

2. Classical finite geometries offer examples of designs either in a direct way (cf.,e.g., Bose 1961, Dembowski 1968) or when we fix within the geometry some properties. A nice example of the latter case is the construction given in Clatworthy 1954. Also non-classical geometries such as the finite weak affine spaces of Sperner are examples of designs.

A weak affine space or shortly an S-space, S, consists of a set P of elements called points and a set L of elements called lines, with an incidence relation I defined in $P \times L$ and a parallelism $\|$ defined in $L \times L$, satisfying the following axioms:

1) Any two distinct points are incident with exactly one line;
2) Any line is incident with the same cardinal number of points, called the order of S;
3) Parallelism is an equivalence relation;
4) For any point-line pair (A, a) there exists exactly one line incident with A and parallel to a.

The notion of an S-space generalizes the notion of an affine space, since affine spaces form a special class of S-spaces. Finite S-spaces, that is S-spaces with a finite number of points, are exactly the structures called 2-(v,k,1) designs with parallelism.

Proper S-spaces may arise from a suitable modification of ordinary affine spaces. A very simple example is the following one. Let $\Sigma(P,L,I,\|)$ be an ordinary affine space of dimension $d > 2$ and order $s \geq 3$. Consider a plane $\alpha$ in $\Sigma$ and two points A and B on $\alpha$. From $\Sigma$, we can obtain an S-space $S(P, L, I', \|)$ by simply changing the incidence relation in the following way: The incidence will be the same as in $\Sigma$ except for the lines which lie in $\alpha$ and pass through A and not through B or pass through B and not through A. For a line of $\alpha$ through A (not AB), the incidence is changed by replacing A with B. Similarly for a line of $\alpha$ through B.

It is easy to see that for the new structure the four postulates for S-spaces are satisfied. The plane $\alpha$ remains desarguesian after the change, but the space is no longer an ordinary affine space since the Veblen axiom is not satisfied. To see this, let r be a line in $\alpha$ through A but not through B (when regarded as a line of $\Sigma$), and let u be a line not in $\alpha$ which passes through A. Then r and u determine a plane $\beta$ in $\Sigma$. Let a and b be two lines of $\beta$ not through A which intersect in a point belonging to neither r nor u . In the S-space, the lines a and b intersect, but not the lines r and u . Thus the Veblen axiom for an ordinary affine space is not satisfied. Clearly the construction can be generalized by exchanging in the above way several pairs of points. Other constructions of S-spaces through modifications of ordinary affine spaces are known (see e.g. Barlotti & Nicoletti 1976).

The methods of descriptive geometry for representing the three-dimensional space suggest natural ways to embed non-desarguesian planes in S-spaces. The construction related to the orthographic projections of Monge is given in Sperner (1960) and arises as a natural product of planes. This construction can be extended also to the high dimensional cases (Barlotti 1965, Herzer 1977). Similar products may also be introduced for the construction of several types of other incidence structures (cf. Glynn 1978, Halder 1979).

3. A spread of a projective geometry (see, e.g. Dembowski 1968) may be very useful in the construction of non-desarguesian planes and of other incidence structures. We recall here the construction of $\Delta$-planes given in Bose & Barlotti 1971, which we shall use in a later section.

Let $\Sigma' = PG(4,q)$ be a finite 4-dimensional space of order q. Let $\Sigma$ and $\Omega$ be two distinct 3-spaces of $\Sigma'$ and denote by $\underline{S}$ a spread of lines of $\Sigma$ . Denote by s the line of $\underline{S}$ contained in the plane $\pi$ intersection of $\Sigma$ and $\Omega$ . The plane to be constructed will be denoted by $\Delta$ and its points and lines will be called $\Delta$-points and $\Delta$-lines to avoid confusing them with elements of $\Sigma'$ which represent them. $\Delta$-points are of three types. $\Delta$-points of type (1) are represented by the planes of $\Sigma'$ which pass through a line of $\underline{S}$ other than s and are not contained in $\Sigma$ . $\Delta$-points of type (2) are represented by planes passing through s and not lying in either $\Sigma$ or $\Omega$ , and $\Delta$-points of type (3) are represented by the points of s.

$\Delta$-lines are also of three types. $\Delta$-lines of type (1) are represented by points of $\Sigma'$ not contained in $\Sigma$ or $\Omega$. $\Delta$-lines of type (2) are represented by planes contained in $\Omega$, not passing through s. There is only one line of type (3) which is represented by s.

The incidence between $\Delta$-points and $\Delta$-lines is defined as follows. A $\Delta$-line of type (2) is incident with a $\Delta$-point of type (1) or (2) if and only if the planes representing them intersect in a line. In every other case a $\Delta$-line is incident with a $\Delta$-point if and only if the element of $\Sigma'$ representing the $\Delta$-line is contained in or contains the element of $\Sigma'$ representing the $\Delta$-point. $\Delta$ is a semi-translation plane of order $q^2$ and the ternary ring of this plane is obtained in Bose & Smith 1971.

The construction of $\Delta$ allows various generalizations. By choosing in a suitable way the dimension of $\Omega$ in PG(2n,p) we obtain other classes of semitranslation planes derived from shear planes of degree greater than 2 over their kernel (Krier 1974). A different generalization of the construction of $\Delta$ has been given in Heft 1971 by taking $\Sigma' = PG(2n,q)$, $\Sigma$ and $\Omega$ to be hyperplanes of $\Sigma'$ and the spread $\underline{S}$ to be a Galois spread of (n-1)-dimensional flats in $\Sigma$. The incidence structure constructed in this way is a divisible design. Heft has also given other constructions of various classes of partial designs using spreads in high dimensional linear spaces and on certain quadrics.

4. Geometrical properties of a curve in a finite plane or of a variety in a space have been used for the construction of designs. Well known examples are in Bose 1958,1961, Bose & Chakravarti 1966, Ray-Chaudhuri 1962, 1965. Other examples are given in Barlotti 1968, Hölz 1981, Pellegrino 1973, 1974, 1977, 1978. We shall recall here briefly the construction of a unital of lines U* (i.e. the dual of a unital arc) in the plane $\Delta$ (Barlotti & Lunardon 1979). The existence

of $U^*$ implies that in $\Delta$ there exists also a unital arc which consists of the points of $\Delta$ which are incident with exactly one line of $U^*$. The construction is analogous to that given in Buekenhout 1976.

Let $\Delta$ be the plane constructed using $\Sigma'$, $\Sigma$, $\Omega$ and s in § 3. Let $\Lambda$ be a hyperplane of $\Sigma'$ which intersects $\Sigma$ and $\Omega$ in two distinct planes denoted $\sigma$ and $\omega$ respectively. Consider an ovoid Q of $\Lambda$ which is tangent to $\sigma$ and $\omega$ in the points P and R respectively. Let p be the line of $\underline{S}$ incident with P and assume that $\Lambda$ is chosen in such a way that p intersects $\Lambda$ only in P. Let V be the intersection of p with $\Omega$ (clearly V is different from P), and denote by C the (3-dimensional) cone of $\Sigma'$ obtained by projecting Q from the point V. Consider in the plane $\Delta$ the set $U^*$ of lines represented in $\Sigma'$ either by the points of C which do not belong to $\Sigma \cup \Omega$, or by the planes of $\Omega$ passing through the line VR. Since $U^*$ consists of $q^3 - q$ lines of type (1) and $q + 1$ lines of type (2), in order to prove that $U^*$ is a unital of lines it is enough to prove that every $\Delta$-point incident with two $\Delta$-lines of $U^*$ is incident with exactly $q + 1$ $\Delta$-lines of $U^*$. This is not difficult and requires only to consider all possible cases (cf. Barlotti & Lunardon 1979).

5. Finite inversive planes are a well known class of designs. Other types of finite Benz planes may be used to construct interesting designs. In Biscarini & Conti 1980 from a Laguerre plane of odd order q is constructed a symmetric design with the same parameters of PG(3,q) (as a structure of points and planes) but not isomorphic to PG(3,q).

Among other incidence structures which in the finite case may be used for the construction of designs we point out the half-planes and vifs. For the theory of these structures see Glynn 1985, where it is also proved how 2-designs can be associated to them.

6. Another geometric way which sometimes leads to the

construction of families of designs may be based on the knowledge of a covering or of a partition of a geometric structure with substructures chosen in a proper way. We may fix two particular families of elements, A and B respectively, as the set of blocks and the set of points of a design, the incidence in the design being given by a particular relation.

We present here one of the nice examples given in Vecchi 1984. The author first proves the following.

Lemma: Consider in $PG(3,q)$ (q odd), a $(q^2+1)$-cap $Q$ and a plane $\alpha$ tangent to $Q$ in a point $V$. Then it is possible to construct in $PG(3,q)\setminus\alpha = AG(3,q)$ a set $D$ of $q$ disjoined $q^2$-caps such that

(i) $Q\setminus\{V\}\in D$

(ii) the caps of $D$ form a partition of the points of $AG(3,q)$.

From this follows the Theorem: Taking as blocks the lines of $PG(3,q)$ (q odd), except those through $V$ and those belonging to $\alpha$, as points the $q^2$-caps of the set $D$, and establishing that a point and a block are incident if and only if the corresponding $q^2$-cap and line meet in two distinct points, we obtain a design with parameters: $v=q$, $b=q^3(q+1)$, $k=(q-1)/2$, $r=q^2(q^2-1)/2$, $\lambda=q^2(q+1)(q-3)/4$.

Other examples based on a covering of a line with a family of sets of points are those considered in Bruck 1969 pp. 428-429 and Cofman 1973. The results are inversive planes and other 3-designs. Finally we will recall two nice constructions obtained recently: in Ebert 1985 it is shown how designs can be obtained by partitioning $PG(3,q)$ into ovoids, and in Kestenband 1986 designs are constructed using properties of $(q+1)$-arcs in $PG(2,q)$ (q odd).

7. When I was invited to present this lecture I gave an indefinite title since at that time I was busy with problems far from Combinatorics and I wanted to remain free to decide. Then I was thinking also of another possibility for this lecture in order to

celebrate the 30-th anniversary of the second birth of a classification of projective planes. However later I thought that it was better to keep the celebration for the 50-th anniversary and I decided for the topic presented above.

Clearly, due to the large variety of types of designs, in order to have a nice classification we have to choose a particular good class of designs, e. g. , the symmetric designs. For those who wish to know everything on the classification of symmetric designs I present here the necessary information: Butler 1984 a,b,1985, Kelly 1983, Piper 1982, 1983, Schulz 1967. Other types of designs may be classified following different principles. For a classification of divisible translation designs cf. Schulz 1985.

## REFERENCES

Barlotti, A. (1965). Alcuni risultati nello studio degli spazi affini generalizzati di Sperner. Rendiconti Sem. Mat. Univ. Padova 35, pp. 18-46.

Barlotti, A. (1968). Procedimenti geometrici per la costruzione di PBIB-disegni. Periodico di Mat. (4) 46 pp. 59-68.

Barlotti, A. (1975). Representation and construction of projective planes and other geometric structures from projective spaces. Jber. Deutsch. Math.-Verein. 77 (1975), pp. 28-38.

Barlotti, A. (1985). Geometric procedure for the construction of designs. Ars Combin. 19-A, pp. 7-14.

Barlotti, A. & Cofman, J. (1974). Finite Sperner spaces constructed from projective and affine spaces. Abh. Math. Sem. Univ. Hamburg 40,pp. 231-241.

Barlotti, A. & Lunardon, G. (1979). Una classe di unitals nei $\Delta$-piani. Riv. Mat. Univ. Parma (4) 5, pp. 781-785.

Barlotti, A. & Nicoletti, G. (1976). On a geometric procedure for the construction of Sperner spaces. Abh. Math. Sem. Univ. Hamburg 45, pp. 251-255.

Barnabei, M. & Brini, A. (1980). A class of PBIB-designs obtained from projective geometries. Discrete Math. 29, pp. 13-17.

Beth, Th., Jungnickel, D. & Lenz, H. (1985). Design Theory. B.I. Wissenschaftsverlag, Mannheim.

Beutelspacher, A. (1982). Einführung in die endliche Geometrie I. B.I. Wissenschaftsverlag, Mannheim.

Beutelspacher, A. (1983). Einführung in die endliche Geometrie II. B.I. Wissenschaftsverlag, Mannheim.

Biscarini, P. & Conti, F. (1980). Piani di Laguerre e disegni simmetrici. Atti Sem. Fis. Univ. Modena 29, pp. 101-107.

Bose, R. C. (1947). The design of experiments, Proc. 34th Indian Sci. Congress, pp. 1-25.

Bose, R. C. (1958). On the application of finite projective geometry for deriving a certain series of balanced Kirkman arrangements. The golden jubilee commemoration volume, Calcutta Math. Soc. pp. 341-354.

Bose, R. C. (1961). Lecture Notes. Dept. of Statistics. Chapell Hill, N.C.

Bose, R. C. & Barlotti, A. (1971). Linear representation of a class of projective planes in a four dimensional projective space. Annali di Mat. Pura e Appl. 88, pp. 9-32.

Bose, R. C. & Chakravarti, I. M. (1966). Hermitian varieties in a finite projective space $PG(N,q^2)$. Canadian J. Math. 18, pp.1161-1182.

Bose, R. C. & Smith, K. J. C. (1971). Ternary rings of a class of linearly representable semi-translation planes. In atti Convegno Geometria Combinatoria, Perugia Università pp. 69-101.

Bruck, R. H. (1969). Construction problems of finite projective planes,in "Combinatorial Mathematics and its applications", pp. 426-514. Univ. of North Carolina Press, Chapel Hill, N.C.

Buekenhout, F. (1976). Existence of unitals in finite translation planes of order $q^2$ with a kernel of order q, Geom. Ded. 5, pp. 189-194.

Butler, N. T. (1984a). On semi-translation blocks. Geom. Dedicata 16, pp. 279-290.

Butler, N.T. (1984b). Symmetric designs of type III. Mitt. Math. Sem. Giessen, pp. 1-24.

Butler, N.T. (1985). Symmetric designs of type V and VI. J. Statist. Plann. Inference 11, pp. 355-361.

Clatworthy, W. H. (1954). A geometrical configuration which is a partially balanced design. Proc. Am. Math. Soc. 5, pp. 47-55.

Cofman, J. (1973). On combinatorics of finite projective spaces. Proc. Intern. Conf. Projective Planes, pp. 59-70, Washington State University Press, Pullman, Wash.

Dembowski, P.(1968). Finite Geometries, Springer-Verlag, Berlin Heidelberg-New York.

Ebert, G. L. (1985). Partitioning projective geometries into caps. Can.J. Math. 37, pp. 1163-1175.

Glynn, D. G. (1978). Finite projective planes and related combinatorial systems, Ph.D. Thesis, Univ. of Adelaide.

Glynn, D.G. (1985). An introduction to half planes, Ars Combin. 19A, pp. 309-342.

Halder, H.-R. (1979). Produktbildung von Inzidenzräumen mit Parallelismus. J. Reine Angew. Math. 305, pp. 82-88.

Hedayat, A. & Kageyama, S. (1980). The family of t-designs - Part I . J. Statist. Plann. Inference 4, pp. 173-212.

Heft, S. M. (1971). Spreads in projective geometry and associated designs. Ph.D. Thesis, Chapel Hill, N.C.

Herzer, A. (1977). Konstruktion von Spernerräumen aus einer vorgegebenen projektiven Ebene. Abh. Math. Sem. Univ. Hamburg 46, pp. 25-54.

Hölz, G. (1981). Construction of designs which contain a unital. Arch.Math. 37, pp. 179-183.

Hughes, D. R. & Piper, F. C. (1985). Design Theory, Cambridge University Press.

Kageyama, S. & Hedayat, A. (1983). The family of t-designs - Part II.J. Statist. Plann. Inference 7, pp. 257-287.

Kelly, G. S. (1983). Symmetric Designs with elations. In "Finite geometries" (N.L. Johnson, M.J. Kallaher, C.T. Long., eds.) Lecture Notes in pure and applied mathematics 82, M. Dekker Inc. pp.263-272.

Kestenband, B. C. (1986). Balanced incomplete block designs on q+1 elements. J. Statist. Plann. Inference 13, pp. 45-50.

Krier, N. (1974). Linear representation of derived shear planes. Boll. UMI (4) 9, pp. 709-720.

Pellegrino, G. (1973). Sulle proprietà della 11-calotta completa di $S_{4,3}$ e su un BIB-disegno ad essa collegato. Bollettino UMI (4) 7, pp. 463-470.

Pellegrino, G. (1974). Costruzione di una classe di schemi di associazione e di PBIB-disegni. Atti Accad. Naz. Modena (6), 16, pp. 5-17.

Pellegrino, G. (1977). Schemi di associazione negli spazi lineari finiti. Rend. di Mat. (4) 9, pp. 645-656.

Pellegrino, G. (1978). t-designs associated with non degenerate conics in a Galois plane of odd order. J. Statist. Plann. Inference 2, pp. 307-312.

Piper, F. (1982). On axial automorphism of symmetric designs. Ann. Discrete Math. 15, pp. 333-340.

Piper, F. (1983). On symmetric designs which admit axial automorphisms in "Finite geometries" (N.L. Johnson, M.J. Kallaher, C.T. Long, eds.), Lectures Notes in pure and applied mathematics 82, M. Dekker Inc., pp. 391-400.

Raghavarao, D. (1971). Constructions and combinatorial problems in design of experiments. J. Wiley & Sons, New York.

Ray-Chaudhuri, D.K. (1962). Applications of the geometry of quadrics for constructing PBIB-designs. Ann. Math. Stat. 33, pp. 1175-1186.

Ray-Chaudhuri, D.K. (1965). Some configurations in finite projective spaces and partially balanced incomplete block designs. Can.

J. Math. 17, pp. 114-123.

Schulz, R.-H. (1967). Über Blockpläne mit transitiver Dilatationsgruppe. M. Zeit. 98, pp. 60-82.

Schulz, R.-H. (1985). On the classification of translation group-divisible designs. Europ. J. Combinatorics 6, pp. 369-374.

Segre, B. (1965). Istituzioni di Geometria Superiore III: Complessi,Reti, Disegni. Roma: Istituto Matematico.

Sperner, E. (1960). Affine Räume mit schwacher Inzidenz und zugehörige algebraische Strukturen. J. Reine Angew. Math. 204, pp. 205-215.

Vecchi, I. (1984). Some results on coverings of Galois spaces with ovoids and related BIB-designs. J. Statist. Plann. Inference 10, pp.219-225.

# *PORTRAIT OF A TYPICAL SUM-FREE SET*

P.J. Cameron

School of Mathematical Sciences
Queen Mary College
University of London
Mile End Road
LONDON E1 4NS

## *INTRODUCTION*

A set S of natural numbers is called <u>sum-free</u> if, for all x, y $\in$ S, we have x + y $\notin$ S.

The history of sum-free sets can be said to date from 1916, when Schur proved that the natural numbers cannot be partitioned into finitely many sum-free sets. They (or their analogues in finite abelian groups), give rise to vertex-transitive triangle-free graphs, yielding applications in Ramsey theory, extremal graph theory, blocking sets in projective planes, and model theory (homogeneous relational structures).

I was first led to the investigations reported here by these applications, which I describe in more detail in the first section of the paper. However, the subject quickly takes on a life of its own. I have chosen this topic for two main reasons. Firstly, the results are often quite surprising and counter-intuitive; I do not believe that a general theory can be built until such a remarkable special case is better understood. Secondly, the methods used in defining and analysing the notion of a "typical" sum-free set are taken from many areas - topology, probability and Hausdorff dimension among them. I think that specialists in these areas may possess tools which could be used effectively on our problems, and vice versa; but we won't find out until we learn one another's language.

A subsidiary reason involves the collection of evidence. Many of the conjectures were formulated (and a few of them proved) on the basis of computation. Indeed, the subject is still at an experimental stage. Moreover, typical computations require a lot of time but little memory, and are best performed on a microcomputer.

Readers are encouraged to try their hand at further data-gathering!

## 1 *ORIGINS AND APPLICATIONS*

### 1.1 *Schur, van der Waerden, Szemerédi and Furstenberg*

I have already mentioned Schur's theorem (1916) that $\mathbb{N}$ cannot be partitioned into finitely many sum-free sets. This was the first result in what is now known as Ramsey theory (excluding Dirichlet's pigeonhole priciple), pre-dating Ramsey by fourteen years. A similar theorem of van der Waerden (1927) also predates Ramsey: if $\mathbb{N}$ is partitioned into finitely many sets, one of them must contain arbitrarily long arithmetic progressions. This led Erdös and Turán (1936) to conjecture that a set of natural numbers of positive upper density must contain arbitrarily long arithmetic progressions.

(The <u>upper</u> and <u>lower</u> <u>density</u> of a subset S of $\mathbb{N}$ are the limit superior and limit inferior, respectively, of $|S \cap \{1,\ldots,n\}|/n$; if they are equal, so that the limit exists, the common value is the <u>density</u> of S).

Progress was initially slow. Roth (1952) proved that a set of positive upper density contains three-term arithmetic progressions. This was improved to four-term progressions by Szemérédi (1969); six years later, he proved the general conjecture. The proof is quite hard. But in 1977, Furstenberg gave a nicer proof using ergodic theory. His methods enriched both combinatorics and ergodic theory. See Petersen (1983) for discussion.

No such result is possible for sum-free sets. The set of all odd numbers is obviously sum-free and has density ½. But the analogy suggests a problem we consider in detail: what can be said about the density of a sum-free set? In particular, does a "typical" sum-free set have a density, and if so, what is it?

### 1.2 *Homogeneous graphs and cyclic graphs.*

We consider finite or countable simple undirected graphs. A graph $\Gamma$ is said to be <u>homogeneous</u> if, whenever $\theta$ is an isomorphism between finite induced subgraphs of $\Gamma$, there is an automorphism of $\Gamma$ which extends $\theta$. Thus "homogeneity of $\Gamma$" means "transitivity of the automorphism group of $\Gamma$ on vertices, edges, non-edges, etc." Clearly this is a very strong condition.

The finite homogeneous graphs were determined by **Gardiner** (1976). The non-trivial countable homogeneous graphs were discovered by Henson (1971) (apart from one found earlier); and Lachlan & Woodrow (1980) proved that the list is complete. The homogeneous graphs are

(i) disjoint unions of m complete graphs of size n, where m, n are finite or countable;

(ii) complements of (i) (complete multipartite graphs);

(iii) the pentagon;

(iv) the line graph of $K_{3,3}$;

(v) for each $n \geq 3$, a graph $G_n$ containing no $K_n$ but having all finite $K_n$ - free graphs as induced subgraphs (these properties and homogeneity characterise it);

(vi) complements of (v);

(vii) the "universal" or "random" graph R (Erdös and Rényi (1963), Rado (1964)).

Types (i) and (ii) are of no interest, while (iii) and (iv) being finite are not relevant here; so we are interested in the graphs $G_n$ ($n \geq 3$) and R. We consider the question: what cyclic automorphism do these graphs have? Henson observed that R and $G_3$ admit cyclic automorphisms, but $G_n$ ($n \geq 4$) does not.

Let $\alpha$ be a cyclic automorphism of a countable graph $\Gamma$. We can number the vertices of $\Gamma$ as $x_n$ ($n \in \mathbf{Z}$) so that $\alpha$ maps $x_n$ to $x_{n+1}$. Let $S = \{ n : n > 0,\ x_n \text{ joined to } x_0\}$. Then S determines $\Gamma$ up to isomorphism: $x_i$ is joined to $x_j$ if and only if $|j - i| \in S$. Furthermore, S determines $\alpha$ up to conjugacy in $\mathrm{Aut}(\Gamma)$. (More precisely, if $\alpha_1$ and $\alpha_2$ are cyclic automorphisms of $\Gamma$, and $S_1$ and $S_2$ the associated subsets of $\mathbb{N}$, then $\alpha_1$ and $\alpha_2$ are conjugate in Aut($\Gamma$) if and only if $S_1 = S_2$).

The case $\Gamma = R$ was considered in detail in Cameron (1984); I recall the results. A subset of $\mathbb{N}$ is called <u>universal</u> if, given any finite sequence $\sigma = (e_1,\ldots,e_n)$ of zeros and ones, there exists m such that, for $1 \leq i \leq n$, $m + i \in S$ if and only if $e_i = 1$. (In other words, if we identify S with its characteristic function, an infinite sequence of zeros and ones, S is universal if and only if S contains every finite sequence). Now the graph $\Gamma(S)$ determined by S is isomorphic to R if and only if S is universal. Moreover, in either of the senses described later in this paper, a typical set is universal. (This includes the assertion that $2^{\aleph_0}$ universal sets exist, so that R admits $2^{\aleph_0}$ non-conjugate cyclic automorphisms).

For cyclic automorphisms of $G_3$, we begin with the observation that $\Gamma(S)$ is triangle-free if and only if S is sum-free. Clearly no sum-free set can be universal: if $k \in S$ and S is sum-free, then no two elements of S can differ by k, and so no finite sequence with ones in positions k apart can occur in S. But this should be the only restriction. So we say that a sum-free set S, with characteristic function S, is <u>*sf*-universal</u> if, given any finite zero-one sequence $\sigma = (e_1,\ldots,e_n)$, either

(i) there exist i, j with $1 \leq i < j \leq n$ with $e_i = e_j = 1$ and $j - i \in S$; or

(ii) $\sigma$ is a subsequence of S.

Now it is not too hard to show that, for any sum-free set S, $\Gamma(S)$ is isomorphic to $G_3$ if and only if S is sf-universal. (This is less trivial than the analogous fact for R). So we must investigate the questions: what do sf-universal sets look like? In particular, is a typical sum-free set sf-universal?

### 1.3 *Finite Applications*

What we did for infinite cyclic groups in the last section

can be re-done for finite cyclic groups.

Let T be a subset of $\mathbb{Z}/(n)$, the cyclic group of order n. Then T is <u>sum-free</u> if x, y $\in$ T implies x + y $\notin$ T. Moreover, we say that T is <u>complete</u> if, for any z $\notin$ T, there exist x, y $\in$ T such that x + y = z; and T is <u>symmetric</u> if x $\in$ T implies - x $\in$ T.

For any subset T of $\mathbb{Z}/(n)$, we define a directed graph with vertex set $\mathbb{Z}/(n)$, by joining x to y if and only if y - x $\in$ T. This graph admits the cyclic group, acting by translation. Moreover, it is loopless if and only if 0 $\notin$ T; undirected if and only if T is symmetric; triangle-free if and only if T is sum-free; and of diameter 2 if and only if T is complete.

Hanson & Seyffarth (1984), investigating a problem of Erdős on critical graphs, were led to ask: what is the smallest valency of a regular triangle-free graph of diameter 2 on n vertices? It is trivial that the least such valency is not less than $\sqrt{n} - 1$. They gave a construction to show that this bound is within a constant factor of being best possible. It suffices to give a small complete symmetric sum-free set in $\mathbb{Z}/(n)$. They showed that, if $n = t^2 + 5t + 2$, then

$$T = \{1, t + 6, 2t + 11, \ldots, t^2 + 5t + 1\} \qquad (1)$$

$$\cup \{t + 3, 2t + 7, 3t + 11, \ldots, t^2 + 4t - 1\}$$

is sum-free, complete, and symmetric, with $|T| = 2t + 1$.

Similar ideas have been effective in other problems. For example, lower bounds for Ramsey numbers for triangles are found by partitioning $\mathbb{Z}/(n)$ (or any finite abelian group) into symmetric sum-free sets: see Hanson (1976). Also such sets give rise to minimal blocking sets in finite Desarguesian projective planes contained in the union of three lines: see Cameron (1985 b). Finally, we refer to the lecture notes by Wallis et al. (1972), of which the part written by Ann Street deals with these questions in much greater detail.

Before leaving this topic, I mention the outcome of a curious question that I asked at the last British Combinatorial Conference. Easy constructions show the existence of complete sum-free sets mod n for all $n \neq 1, 3, 7$. At the time, however, all examples known to me were symmetric, and I asked whether this was necessarily so. The smallest counter-example is surprisingly large, with n = 36, and was sent to me by Neil Calkin. Subsequently, Dominic Welsh found a randomised algorithm which produced a plethora of such examples. However, some problems remain:

(i) Is there a number $n_0$, such that non-symmetric complete sum-free sets exist in $\mathbb{Z}/(n)$ for all $n \geq n_0$?

(ii) Is there an upper bound, less than 1, for $|T'|/|T|$, where T is a complete sum-free subset of $\mathbb{Z}/(n)$ and $T' = \{x \in T: -x \notin T\}$?

## 2 *THE STRUCTURE OF* $2^{\mathbb{N}}$

### 2.1 *As a Metric Space*

In order to describe a typical object of a set, some structure must be put on the set, such that a class of naturally-occurring "large" subsets exist. Two traditional ways of doing this involve Baire category (in complete metric spaces) and measure. We refer to Oxtoby (1980) for a discussion of these concepts and the relationship between them.

As described in Section 1.2, we identify any set S of natural numbers with its characteristic function, a sequence s of zeros and ones with s(n) = 1 if and only if $n \in S$. It is natural to regard a sequence as being generated term by term, and to think of two sequences as being close together if we have to wait a long time before being able to distinguish them. Formally, set

$$d(s,t) = \begin{cases} 0 \text{ if } s = t \\ 1/2^n \text{ if } s(i) = t(i) \text{ for } i \leq n \text{ but} \\ s(n+1) \neq t(n+1). \end{cases} \qquad (2)$$

It is readily checked that d is a metric on $2^{\mathbb{N}}$. Indeed, it satisfies a strengthened form of the triangle inequality known as the <u>ultrametric</u>

inequality:

$$d(s, u) \leq \max (d(s,t), d(t,u)). \tag{3}$$

It is also easy to see that $(2^{\mathbf{N}},d)$ is a complete metric space. (A sequence $(s_n)$ of sequences converges to a sequence s if and only if for all i, $s_n(i) = s(i)$ for all sufficiently large n).

Let $2^*$ denote the set of finite sequences of zeros and ones. Any ball (open or closed) in $2^{\mathbf{N}}$ consists of all sequences having $\sigma$ as an initial segment, for some $\sigma \in 2^*$. Using this, it is apparent that

(i) $U \subseteq 2^{\mathbf{N}}$ is open if and only if any $s \in U$ has an initial segment $\sigma$ with the property that every sequence with initial segment $\sigma$ is in U. (We say that such a U is finitely determined).

(ii) $U \subseteq 2^{\mathbf{N}}$ is dense if and only if every finite sequence is an initial segment of some member of U. (We say that such a U is always reachable).

A set is residual if it contains a countable intersection of open dense sets. The Baire category theorem asserts that, in a complete metric space, any residual set is non-empty. The usual interpretation is that residual sets are "large"; for example, it follows immediately that the intersection of countably many residual sets is non-empty (and in fact is residual).

In our case, the proof is a straightforward exercise. Let U be residual; we may assume that $U = \bigcap_{n=1}^{\infty} U_n$, where $U_n$ is open and dense. Define a sequence $(\sigma_n)$ of elements of $2^*$ as follows: $\sigma_0 = \emptyset$. Given $\sigma_n$, it has a continuation $s_{n+1}$ in the (always reachable) set $U_{n+1}$, and there is an initial segment $\sigma_{n+1}$ of $s_{n+1}$ all of whose continuations lie in $U_{n+1}$ ; we may suppose that $\sigma_{n+1}$ properly extends $\sigma_n$.

Then the sequence $(\sigma_n)$ defines an infinite sequence s, and $s \in \bigcap_{n=1}^{\infty} U_n = U$.

The proof can be modified to show that residual sets are dense and have cardinality $2^{\aleph_0}$. To show denseness, let $\sigma$ be any finite sequence, and start the construction with $\sigma_0 = \sigma$. For the cardinality result, "code" elements of $2^{\mathbf{N}}$ into the construction by inserting one extra digit at the end of each $\sigma_n$ when it is constructed.

To illustrate the concept, we show that the set *Uni* of universal sequences is residual. We have

$$Uni = \bigcap_{\sigma \in 2^*} U(\sigma), \tag{4}$$

where $U(\sigma)$ is the set of sequences having $\sigma$ as a subsequence; and it is clear that $U(\sigma)$ is both finitely determined and always reachable. In particular this demonstrates that the "random" graph R has $2^{\aleph_0}$ non-conjugate cyclic automorphisms, as noted in Section 1.2.

### 2.2 *As a Probability Space*

The treatment in this section will be less formal, since these are two good intuitive ways of describing a probability measure on $2^{\mathbf{N}}$, both giving the same result:

(i) a random sequence is chosen by selecting its terms independently, with probability ½ for each value of the nth term. In other words, it is generated by a countable sequence of tosses of a fair coin.

(ii) Map $2^{\mathbf{N}}$ to the unit interval by regarding each sequence as a binary decimal. This map is one-to-one outside a countable (and hence null) subset. Now use Lebesgue measure on [0,1].

The "large" sets, analogous to the residual sets of the last section, are the sets of measure 1. Thus, for example, any countable intersection of sets of measure 1 has measure 1, and any set of measure 1 has cardinality $2^{\aleph_0}$.

I illustrate with the analogue of the earlier result: the set

*Uni* has measure 1. For we can write, as before, $Uni = \bigcap_{n=1}^{\infty} U(\sigma)$. If $\sigma$ has length k, then the set of sequences which don't have a copy of $\sigma$ beginning at a given position has measure $1 - 1/2^k$; so the set having no copy of $\sigma$ beginning in any of positions 1, k + 1, 2k + 1,...(m - 1)k + 1 has measure $(1 - 1/2^k)^m$, since these events are independent. So the intersection of these sets for all m is null and contains the complement of $U(\sigma)$.

Don't think, though, that measure and category always give the same description of a typical set. Here are two naturally-occurring "large" sets (in the two senses) which are disjoint. Recall the definitions of upper and lower density in Section 1.1. The Strong Law of Large Numbers asserts that the set

$$\{S \subseteq \mathbb{N} : S \text{ has density } \tfrac{1}{2}\} \tag{5}$$

has measure 1. On the other hand,

$$\{S \subseteq \mathbb{N} : S \text{ has upper density 1 and lower density 0}\} \tag{6}$$

is residual. To see this, it is enough to show that

$$V_\varepsilon = \{S \subseteq \mathbb{N} : S \text{ has lower density} \le \varepsilon\} \tag{7}$$

is residual. This follows from the fact that $V_\varepsilon = \bigcap_{n=1}^{\infty} V_\varepsilon(n)$, where $V_\varepsilon(n) = \{S \subseteq \mathbb{N} : \exists m \ge n \text{ such that } |S \cap \{1,\ldots m\}|/m \le \varepsilon + 1/n\}$ is open and dense - this is readily checked.

### 2.3 *Fractal Dimension*

Most sets of interest, such as the set of all sum-free sets, are "small" in both senses, so we need different methods for them. I will concentrate on a technique involving a natural bijection with $2^{\mathbb{N}}$; but there is a different approach worth considering, involving an interaction of measure and topology.

Let (X,d) be any metric space. For a subset Y of X, and real numbers $s \ge 0$ and $\delta > 0$, we define

$$\mu^s_\delta(Y) = \inf_{\mathcal{U}} \sum_{U \in \mathcal{U}} (\text{diam } U)^s; \tag{8}$$

where the <u>diameter</u> diam U of a set U is $\sup_{x,y \in U} d(x,y)$ and the infimum is taken over all coverings of U by countably many sets of diameter $\delta$ or less. Now $\mu^s_\delta(Y)$ increases as $\delta$ decreases, since the infimum is taken over a smaller set;

$$\mu^s(Y) = \lim_{\delta \to 0} \mu^s_\delta(Y) \tag{9}$$

exists (possibly $\infty$). The function $\mu^s$ is an outer measure on the power set of X, and defines a measure on a $\sigma$-algebra of subsets of X in a standard way; this is known as s-<u>dimensional</u> <u>Hausdorff</u> <u>measure</u>. See Rogers (1970), Falconer (1985) for further discussion. It can furthermore be shown that, given Y, there is a unique number $s_0$ such that

$$\mu^s(Y) = \begin{cases} \infty \text{ if } s < s_0 \\ 0 \text{ if } s > s_0. \end{cases} \tag{10}$$

The number $s_0$ is called the <u>fractal</u> or <u>Hausdorff</u> <u>dimension</u> of Y, written dim (Y).

Now we consider the case where $X = 2^{\mathbb{N}}$, with the metric of Section 2.1. Clearly a set U of diameter $\delta$ is contained in a closed ball of radius $\delta$; the ultrametric inequality shows that the diameter of a closed ball of radius $\delta$ is equal to $\delta$ also. Thus, we may assume that all sets U occurring in any covering in the definition are closed balls.

It is not hard to see that $\dim(2^{\mathbb{N}}) = 1$, and that 1-dimensional Hausdorff measure coincides with the measure of Section 2.2 (that is, Lebesgue measure). Moreover, for any s, the $\mu^s$-measurable sets include all Borel sets, and hence all sets of interest to us. We now consider a few examples. Let *Sf* be the set of all sum-free sets; *Odd*, the set of all sets of odd numbers; and *Fib* the set of all sets which contains no two consecutive members. (The connection of the last of these with Fibonacci will appear soon). It is an exercise to show that

$$\dim\,(Odd) = \tfrac{1}{2} \tag{11}$$

and $$\dim\,(Fib) = \log_2 \Phi. \tag{12}$$

where $\Phi = \tfrac{1}{2}(1 + \sqrt{5})$. The dimension of *Sf* is more problematical, but it is at least ½, since it contains *Odd*.

For any subset X of $2^{\mathbf{N}}$, let $f_X(n)$ be the number of sequences of length n which occur as initial subsequences of members of X.

*Lemma 3.1* $\dim(X) \leq \log_2 (\liminf\, (f_X(n)^{1/n}))$.
This is clear from the fact that X can be covered by $f_X(n)$ closed balls each of diameter $\tfrac{1}{2}^n$. We have $f_{Odd}(n) = 2^{[n/2]}$ and $f_{Fib}(n) = F_{n+1}$ (the $(n + 1)^{st}$ Fibonacci number), so equality holds in (3.1) for both of these sets. We cannot expect equality to hold in general. (For example, if X is the set of sequences with only finitely many ones, then X is countable, and so dim(X) = 0; but $f_X(n) = 2^n$ for all n). But I have no example of a perfect set (see next section) for which strict inequality holds.

What is dim (*Sf*)? We have noted the lower bound ½; so, if we could show that $\liminf\, (f_{Sf}(n)^{1/n}) = \sqrt{2}$, then $\dim\,(Sf) = \tfrac{1}{2}$ would follow from (3.1). Numerical evidence supports this guess. I have computed $f_{Sf}(n)$ for all $n \leq 45$; the values of $(f_{Sf}(n))^{1/n}$ are decreasing and the last calculated value is 1.49997... (from $f_{Sf}(45) = 83902379$). More significantly, consider $f_{Sf}(n)/2^{\frac{1}{2}n}$, an "estimate" for $\mu^{\frac{1}{2}}(Sf)$. Apart from a parity effect (whereby, if a smooth curve is fitted, terms with odd n lie above it and those with even n below), this quantity increases until n is about 39 or 40, and then begins to decrease. To establish the dimension of *Sf*, we only need to know that, for some subsequence of values, it does not increase exponentially! The numerical results suggest a stronger conjecture, according to which *Sf* has finite ½-dimensional measure. If so, we would have the surprising conclusion that (measured in this way) a positive proportion of the sets in *Sf* lie in *Odd*.

Another interesting set is *3Pf*, the set of all sets containing no 3-term arithmetic progression. This set is much easier to deal with. For any translate of a finite 3-term-progression-free set has the same property, and so, with $f(n) = f_{3Pf}(n)$, we have

$$f(m + n) \leq f(m)\, f(n), \qquad (13)$$

so $\lim (f(n)^{1/n})$ exists and does not exceed any value of $f(n)^{1/n}$. Thus upper bounds for the dimension can be found by computation. But I conjecture that $\lim (f(n)^{1/n}) = 1$ (and so dim (*3Pf*) = 0).

## *PERFECT SUBSETS OF* $2^{\mathbf{N}}$

### 3.1 *A Bijection*

A prefix of a sequence s is an initial subsequence of s; a prefix of a subset X of $2^{\mathbf{N}}$ is a prefix of some member of X. Now it is readily checked that s is a limit point of X if and only if every prefix of s is a prefix of X. Thus, X is closed if and only if any sequence, all of whose prefixes are prefixes of X, is in X; in other words, membership of X is determined by conditions involving prefixes. Thus *Odd, Fib, Sf, 3Pf* are all closed. Moreover, X is perfect (equal to its set of limit points) if and only if it is closed and any prefix of X is a prefix of at least two members of X. Again, our examples all satisfy this condition.

Any perfect subset of $2^{\mathbf{N}}$ is homeomorphic to $2^{\mathbf{N}}$ (with the induced topology); but the explanation of this will suggest a better description of the topology. Also, induced measure is of no use, since X may be null.

We construct, for any perfect set X, a bijection from $2^{\mathbf{N}}$ to X. We regard this bijection as being computed by a "machine" $M_X$ which takes an element of $2^{\mathbf{N}}$ as input and produces an element of X as output. Think of $M_X$ as having read and write heads which are positioned over one way infinite tapes: these heads can move only in the positive direction. However, $M_X$ is also equipped with an "oracle", which can scan what has been written on the output tape (a prefix of X) and answer the question: which of the two possible extensions $\sigma 0$ and $\sigma 1$ of $\sigma$ is a prefix of X? (at least one must be).

The machine repeats the following step ad infinitum:

(i) Ask the oracle which of $\sigma 0$ and $\sigma 1$ is a prefix of X.

(ii) If only one of $\sigma 0$, $\sigma 1$ is a prefix, write the appropriate symbol and advance the write head.

(iii) If both are prefixes, copy one symbol from input to output and advance both heads.

The closure of X ensures that the output sequence is in X. The perfectness of X ensures that step 3 is used infinitely often, so every term of the input is read. It follows that the function computed by $M_X$ is a bijection from $2^{\mathbb{N}}$ to X, as claimed.

We need an oracle in general since there are $2^{\aleph_0}$ perfect sets but only countably many deterministic machines. But it may be interesting to investigate the relation between recursion-theoretic properties of $M_X$ and the structure of X. For example, I conjecture that, if the oracle for $M_X$ can be simulated by a finite-state machine, then equality holds in Lemma 3.1. (This hypothesis is satisfied for *Odd* and *Fib*. In each case, the oracle only needs to remember one bit of information about the output sequence: for *Odd*, whether the write head is at an even or odd numbered position; for *Fib*, the last symbol written. The oracle for *Sf* can be can be simulated by a Turing machine, but not by a finite-state machine. This is intuitively clear, but we'll see a good reason for it later).

Let $e_X$ be the function defined as follows: $e_X(n)$ is the maximum position the write head can attain when the read head is in position n. Linear growth of $e_X$ seems to be indirectly related to positive Hausdorff dimension. Certainly, there are at least $2^n$ different prefixes of X of length $e_X(n)$, that is, $f_X(e_X(n)) \geq 2^n$; so if $e_X(n) \leq cn$, then the right-hand side of 3.1 is at least 1/c. Unfortunately, the inequality goes the wrong way! But a finite-state oracle should also ensure linear growth of $e_X$.

It is easy to see that $e_{Odd}(n) = e_{Fib}(n) = 2n$. What about $e_{Sf}(n)$?

*Proposition 3.1* $cn^{3/2} \leq e_{Sf}(n) \leq \frac{1}{2}n(n + 3)$.

*Proof.* The upper bound is trivial: if n terms of input have been read, the output contains at most n ones, hence at most $\frac{1}{2}n(n + 1)$ positions are sums of pairs of positions containing ones: thus case 2 of the algorithm applies at most $\frac{1}{2}n(n + 1)$ times. For the lower bound we use Hanson & Seyffarth's example (see Section 1.3) of a complete sumfree set T mod $t^2 + 5t + 2$ with $|T| = 2t + 1$. Consider the input which generates r cycles of T as output. The output has length $r(t^2 + 5t + 2)$. Since T is complete, in all cycles after the first the input consists entirely of ones, so has length at most

$$(t^2 + 5t + 2) + (r - 1)(2t + 1). \qquad (14)$$

Choosing $r \sim t$ gives the result. I have computed $e_{Sf}(n)$ for $n \leq 21$; the last value is 105.

Our earlier claim that a perfect subset of $2^{\mathbf{N}}$ is homeomorphic to $2^{\mathbf{N}}$ is true because the bijection from $2^{\mathbf{N}}$ to X defined by $M_X$ is a homeomorphism. (It is continuous, since it satisfies $d(f(x)\ f(y)) \leq d(x, y)$; since $2^{\mathbf{N}}$ is compact and X is Hausdorff, it is also open). However, it is more convenient to define a new metric on X so that the bijection is an isometry - that is, define the distance between two members of X to be the distance (in $2^{\mathbf{N}}$) between the input sequences generating them. Similarly, we define a measure on X so that the bijection is an isomorphism of measure spaces.

The measure has a simple informal description, which I give for *Sf.* Choose a random member s of *Sf* as follows. Consider the natural numbers in turn. While considering n, if $n = x + y$ with $s(x) = s(y) = 1$, then $s(n) = 0$; otherwise, decide the value of s(n) by the toss of a fair coin. A similar description holds for any perfect set.

### 3.2 *Sf - Universality is residual*

In the sense of category, a typical sumfree set is sf-universal:

*Proposition 3.3* The set of sf-universal sequences in residual in *Sf*.

*Proof* As usual, we write this set as $\bigcap_{\sigma\in 2^*} U(\sigma)$, where $U(\sigma) = \{s \in sf$: either $\sigma$ is a subsequence of s, or there exist i,j with $s(j - i) = \sigma(i) = \sigma(j) = 1\}$ and show that $U(\sigma)$ is finitely determined and always reachable. The first is (as usual) obvious. So let T be any prefix of Sf. If there exist i, j with $T(j - i) = \sigma(i) = \sigma(j) = 1$, then any member of *Sf* with prefix T is in $U(\sigma)$; so suppose not. Extend T with zeros to a sequence T′ which is at least twice as long as T, and of length greater than n. Then concatenate T′ with $\sigma$. The result T" is still a prefix of *Sf*. For, if $T''(i) = T''(j) = T''(i + j) = 1$ then, without loss of generality, $i \leq \text{length}(T)$ and $j \geq \text{length}(T')$, so that $T(i) = 1 = \sigma(j - m) = \sigma(i + j - m)$, where $m = \text{length}(T')$, contrary to the case assumption. As noted earlier, this shows that Henson's graph $G_3$ admits cyclic automorphisms, a fact observed by Henson. Indeed, Henson showed the existence of cyclic automorphisms for which consecutive vertices are adjacent, and hence of (rather special) Hamiltonian cycles in $G_3$. Using the Proposition, we can improve this result:

*Proposition 3.4.* Let $v_0, v_1, \ldots, v_n$ be vertices of $G_3$, and suppose that the map $\theta$: $v_i \to v_{i+1}$ $(i = 0,\ldots,n-1)$ is an isomorphism of induced subgraphs. Then there is a cyclic automorphism of $G_3$ extending $\theta$.

*Proof.* Set $\sigma(i) = 1$ if $v_0$ is joined to $v_i$, 0 otherwise, for $i = 0,\ldots n - 1$. Then $\sigma$ is a prefix of *Sf* by assumption, and hence (by Proposition 3.3) a prefix of an sf-universal sequence. Hence there is a graph isomorphic to $G_3$, with a cyclic automorphism $w_i \to w_{i+1}$ $(i \in \mathbf{Z})$, such that the induced subgraph on $\{w_0,\ldots,w_n\}$ is isomorphic to that on $\{v_0,\ldots,v_n\}$. The result follows by homogeneity.

Henson's Hamiltonian cycle is obtained by choosing $n = 2$ and $\{v_0, v_1\}$ an edge.

It is not difficult to show that a residual subset of *Sf* consists of sequences with lower density zero. In particular, many sf-universal sequences have lower density zero. This is not surprising; the implicit construction in Proportion 3.3 introduces long blocks of zeros. But results in the other direction are lacking:

*Problem*

(i) Is it true that the set of sequences in Sf with density zero is residual?

(ii) Prove or refute the conjecture that any sf-universal sequence has density zero.

An affirmative answer to (ii) would settle (i), and would also provide some replacement for the lack of a "density" version of Schur's theorem, discussed in Section 1.

### 3.3 *Generalities about Measure*

By contrast, the structure of *Sf* as measure space is much more interesting. In particular, sf-universality is not typical. The next chapter is devoted to this. By way of introduction, this section contains some general considerations.

A natural question is the extent to which a version of the Strong Law of Large Numbers holds. It cannot generalise in the most obvious way. For example, it is easy to define a perfect set X, all members of which have density zero. (In fact, by Roth's theorem, this is true of *3Pf*). Also, consider the (disjoint) union of the power sets of $2\mathbf{N} + 1$ and of $4\mathbf{N}$. It is easily checked that a random member of this set has probability $^6/_7$ of lying in the first set and, conditional on this, almost surely has density $\frac{1}{4}$; almost all members of the second set have density $\frac{1}{8}$. However, it is an easy exercise to show that, if the expansion function $e_X(n)$ is bounded above by a linear function of n, then the lower density of members of X is almost surely bounded away from 0. In some cases, such as *Sf*, one can also give upper bounds on density

*Proposition 3.6.* A random member of *Sf* almost surely has upper density at most $^1/_3$.

*Proof.* By the Strong Law, given $\epsilon > 0$, at least $1 - \epsilon$ of the inputs of large length n have between $(\frac{1}{2} - \epsilon)n$ and $(\frac{1}{2} + \epsilon)n$ ones, the first of which occurs not later than position $\epsilon n$. But an input with this

**property generates an output of length at least $(^3/_2 - 2\epsilon)n$; for the n bits of input are copied to the output, and if the first one occurs in position m, then at least $(½ - \epsilon)n - m$ positions have the form $x + m$, where the output has a one in position x. Thus the density of the output is at most $(½ + \epsilon)/(^3/_2 - 2\epsilon)$, and the limit superior is at most $^1/_3$.**

**It is a simple exercise to prove that random members of *Odd* and *Fib* almost surely have density $^1/_4$ and $^1/_3$ respectively. As explained in the next chapter, I believe that no such simple result holds for *Sf*; the density of a random member of *Sf* has a more elaborate distribution.**

## 4 *SF AS A MEASURE SPACE*

### 4.1 *The Probability of Odd*

**By analogy with $2^{\mathbf{N}}$ and the fact that sf-universality is residual, it might be expected that this property also has measure 1 in *Sf*. I expected this to hold - indeed, I hoped to construct cyclic automorphisms of $G_3$ by proving it - and was surprised by the following result.**

***Theorem 4.1.*** **The probability p that a random sum-free set consists entirely of odd numbers satisfies**

$$0.21759\ldots \leq p \leq 0.21860\ldots \qquad (15)$$

**The method of proof is given in Cameron (1985), though the specific bounds given above require some refinements. An outline of the method follows. An obvious upper bound for p is the quantity $p_n$, the probability that a member s of *Sf* satisfies $s(2i) = 0$ for all $i \leq n$. This can be calculated directly by summing the measures of the closed balls defined by the $2^n$ prefixes of *Sf* satisfying this condition. In fact it is possible to calculate $p_n$ from a smaller sum involving only $3^{[½n]}$ terms. This has been done for $n \leq 33$; it turns out that $P_{33} = 16142623428197322246/2^{66} = 0.21877\ldots$ Moreover, $p_n - p$ can be bounded below, using ideas like those in the next paragraph, giving the quoted upper bound for p.**

To produce a lower bound for p, we must bound $p_n - p$ above. In fact there is an exponential upper bound. Using the most simple-minded bound, the main conclusion (viz., that $p > 0$) follows once the value of $p_{12}$ has been calculated; this task would be almost within the reach of hand calculation. But the bound can be sharpened so as to yield the quoted lower bound for p (using, again, the value of $p_{33}$).

The theorem shows that the event "s is sf-universal" does not have measure 1. (No member of *Odd* can be sf-universal, for if s is sf-universal, then s contains the subsequence 00, and so there is an odd number n with $s(n) = 0$; but then s contains the sequence consisting of two ones separated by n zeros, and $s \notin$ *Odd*. Alternatively, if S is any set of odd numbers, then the cyclic graph $\Gamma(S)$ is bipartite - the even and odd integers form a bipartition - and $\Gamma(S)$ cannot be the "universal" triangle-free graph $G_3$).

What kind of sequences account for the remaining measure of *Sf*?

## 4.2 *The Main Conjecture*

If T is a sumfree set mod n, let S(n, T) denote $\{x \in \mathbb{N} \mid x \pmod{n} \in T\}$, and E(n, T) the set of subsets of S(n, T) (or the event that a random member of *Sf* lies in this set). Thus *Odd* = E(2, {1}). Perhaps other sets of this form account for the missing measure.

*Proposition 4.2.*

(i) If T is sumfree mod n but not complete then E(n, T) is a null set.

(ii) If $T_1$ and $T_2$ are sumfree mod $n_1$, $n_2$, respectively and $S(n_1, T_1) \neq S(n_2, T_2)$ then $E(n_1, T_1) \cap E(n_2, T_2)$ is a null set.

The proof is straightforward. If T is sumfree mod n and $x \pmod{n} \notin T + T$, then to ensure that a random sumfree set is in $E(n, T)$, we must throw tails at each position $kn + x$ $(k \in \mathbb{N})$. Also $E(n_1, T_1) \cap E(n_2, T_2) = E(n_1 n_2, T_3)$, where $T_3$ is sumfree mod $n_1 n_2$ but not complete.

*Main Conjecture.*

(i) If $T$ is complete sum-free mod $n$, then $p(n, T) > 0$.

(ii) Let $S(n_1, T_1), S(n_2, T_2), \ldots$ be a list of all the distinct sets $S(n, T)$ where $T$ is complete sumfree mod $n$. Then $p(\bigcup E(n_i, T_i,)) = 1$. Equivalently, $\Sigma\, p(E(n_i, T_i)) = 1$.

I have no direct evidence for the conjecture; but this is an experimental science, and there is some indirect or supporting evidence.

In view of Theorem 4.1, it is natural to look for a dissection of *Sf* into countably many pieces of which *Odd* is one. My first guess was that, with probability 1, a random sumfree set contains only finitely many even numbers. This now seems unlikely. Computation suggests that the event "2 is the only even number in S" has measure 0 (though I cannot prove even this). But the intuitive reason for Theorem 4.1. is that, if a prefix of S contains no even numbers, then the odd numbers will be fairly "randomly" distributed, and there is a very high probability that the next even number will be the sum of two odd numbers in S, and hence excluded. This argument works for $E(n, T)$ whenever $T$ is complete sum-free mod $n$. One can show that, if $p$ is the measure of $E(n, T)$, and $p_k$ the probability of the event that the prefix of S of length $kn$ is contained in $S(n, T)$, (that is, if $x \leq kn$ and $x \pmod n \notin T$, then $s(x) = 0$), then

$$p_k - p \leq ac^k, \tag{16}$$

where a and c are constants and $c < 1$.

Hence a proof of part (i) of the conjecture in a particular case involves computing $p_k$ for larger and larger values of k until $p_k > ac^k$.

As in Theorem 4.1., the length of the calculation of $p_k$ grows exponentially with k. I have not succeeded in verifying even the next cases of the conjecture, viz. n = 5, T = {1, 4} or T = {2, 3}. The numerical evidence suggests, however, that the assertion is true in these cases.

Note, incidentally, that the events E(5, {1, 4}) and E(5, {2, 3}) have the same probability. For the bijection on $\mathbb{N}$ defined by

$$5k + 1 \to 5k + 3$$
$$5k + 2 \to 5k + 1$$
$$5k + 3 \to 5k + 4$$
$$5k + 4 \to 5k + 2$$
$$5k + 5 \to 5k + 5$$

maps S(5 {2, 3}) onto S(5, {1, 4}), and pairwise sums from the first set to pairwise sums from the second. The value of $p_{10}$ for each of these cases is about 0.025.

### 4.3 *Densities*

One consequence of the Main Conjecture would be the failure of the Strong Law of Large Numbers for *Sf*. This is because, within any set E(n, T) of positive measure, a law of large numbers holds, but the resulting densities can vary from set to set.

*Proposition 4.3.* Let T be sum-free mod n, and suppose that E(n, T) has positive measure. Then, conditioned on E(n, T), almost every sequence has density |T|/2n.

In this sense, within E(n, T), the members of S(n, T) behave as if they were independent, with probability ½ for either outcome. It is worth noting that they are not really independent or equiprobable. For example, it is an exercise to show that, conditioned on *Odd*, the event s(1) = 0 has probability ⅛. (Said otherwise, this means that tails on the first throw reduces the probability of *Odd* by a factor ¼, while heads increases it by a factor $^7/_4$). More generally, conditioned on *Odd*, the event s(1) = s(3) = ... = s(2k - 1) = 0 has probability $\frac{1}{8}^{k}$.

So the proof of Proposition 4.3 requires a version of the Strong Law of Large Numbers for non-independent variables. The form I was able to establish goes roughly as follows: let $X_1, X_2, \ldots$ be $\{0, 1\}$ - random variables. Assume that

(i) As $m \to \infty$, $p(X_m = 0) \to \frac{1}{2}$, not too slowly;

(ii) For any $(e_1, e_2, \ldots, e_m) \in \{0, 1\}^m$

$$p(X_1 = e_1, \ldots, X_m = e_m) \leq 2^{-m}f(m),$$

where $f(m)$ doesn't grow too rapidly.

Then $(\sum_{i=1}^{n} X_i)/n \to \frac{1}{2}$ almost surely. (17)

In the application, we let $x_i$ be the term of a random member of *Sf* in position j, where j is the $i^{th}$ member of S(n, T), and use conditional probabilities in E(n, T). As to (i), $|p(x_m = 0) - \frac{1}{2}|$ is exponentially small - this follows from the inequation (16) of the preceding section. Also, it is straightforward that

$$p(X_1 = e_1 \ldots, X_m = e_m) \leq 2^{-m}/p(E(n, T)). \quad (18)$$

(Our remark about *Odd* indicates that no analogous lower bound is possible; fortunately, it is not needed).

Thus a consequence of the Main Conjecture is the following conjecture about densities.

*Conjecture.* There is a countable set $D = \{d_1, d_2, \ldots \}$ such that, if s is a random member of *Sf*, then

(i) for any i, $p(s \text{ has density } d_i) > 0$;

(ii) $p(s \text{ has a density, which is in } D) = 1$.

**In other words, there is a countable spectrum of densities of sumfree sets. We can attempt to observe this spectrum experimentally, by designing a "spectroscope" along the following lines. Choose prefixes of Sf of some reasonable length, calculate their densities, and work out the distribution. As always, there is an uncertainty principle at work: the sharper we want the spectral lines to be, the longer we must calculate. Also, we can choose between exhaustive search or Monte Carlo technique. I haven't tried the latter, but have performed exhaustive search up to n = 41. (Only a small modification of the program for calculating $f_{Sf}(n)$ is required). There is, as expected, a strong peak corresponding to density ¼ (from *Odd*), but the resolution is not good enough to separate this from the weaker hypothesised spectral line at $^1/_5$ (from E(5, {1, 4})) and E(5, {2, 3})).**

**The figures do contribute some support for the conjecture. If the distribution were unimodal and behaved in accordance with the Central Limit Theorem, the variance would be $O(1/\sqrt{n})$. In fact, the ratio of the variance to $1/\sqrt{n}$ decreases until about n = 19 and then begins to increase. (The conjecture would imply that the variance tends to a non-zero limit).**

**The main conjecture would imply that the spectrum is the set {|T|/2n : T complete sumfree mod n}. What does this set look like? I mentioned Neil Calkin's listing of complete sumfree sets mod n for n ≤ 36. This suggests that $^1/_6$ is the largest limit point of the spectrum; indeed, the only known values greater than $^1/_6$ are of the form (k + 1)/2(3k + 2) for k = 0, 1, 2, ..., resulting from examples with n = 3k + 2, T = {k + 1, k + 2, ..., 2k + 1} or equivalent sets. ($T_1$ and $T_2$ (mod n) are equivalent if $T_2 = dT_1$ for some d coprime to n). Many other values and limit points follow from explicit constructions (for example, the examples of Hanson & Seyffarth show that 0 is a limit point).**

## *5 PERIODICITY OR CHAOS?*

**In this chapter I will describe a strange question which arose naturally from investigations of the main conjecture.**

**Since the machine M = $M_{Sf}$ defines a bijection from $2^{\mathbf{N}}$ to**

*Sf*, we can consider the reverse action: what input sequence generates a prescribed output? It is easy to see, for example, that if the output is the characteristic function of S(n, T), where T is complete sum-free mod n, then the input is ultimately 1 (that is, all but finitely many terms of the input are 1). Moreover, the heuristic argument for part (1) of the main conjecture might appear to work for the power set of any set generated by an ultimately - 1 input. This is not completely true: the set $(S = \{1\} \cup \{x : x \equiv 2 \pmod 3),\ x \geq 2\}$ is generated by the input 100101111 ..., but the power set of S has measure 0. But at best it seems that this is a good place to look for counter-examples. In other words, can an ultimately - 1 input generate an output which isn't ultimately periodic? (A sequence s is <u>ultimately periodic</u> if there exist m, p with $p \geq 0$ such that $s(x + p) = s(x)$ for all $x \geq m$). More generally,

*Problem.* If the input to $M_{Sf}$ is ultimately periodic, is the output necessarily ultimately periodic?

Note first that, if *Sf* is replaced by a perfect set X for which the oracle can be simulated by a finite state machine, the answer is "yes", and indeed, the output period would be bounded by a function of the input period. No such bound holds for *Sf*. (For example, the set $S(3k + 2, \{k + 1,..., 2k + 1\})$ of the preceding section is generated by an input with k zeros followed by all ones). This demonstrates that *Sf* doesn't have a finite-state oracle.

Next, the converse question has an affirmative answer:

*Proposition 5.1.* If the output from $M_{Sf}$ is ultimately periodic, then the input must also be ultimately periodic.

*Proof.* Suppose that $os(x + p) = os(x)$ for all $x \geq m$. Now suppose that $x \geq 2m + 2p$. If $x = y + z$ with $os(y) = os(z) = 1$, then at least one of y and z (y, say) is m + p or greater; so $os(y - p) = os(y) = 1$, and $(x - p) = (y - p) + z$. Similarly conversely. Thus, in any cycle of os after 2m + 2p, calls to the input sequence occur in the same places, and so the input must be periodic.

A very similar argument provides us with a practical test

for periodicity. Suppose that the input sequence is ultimately periodic. Suppose further that the output sequence is periodic with period p from m to 2m + 2p - 1 (that is, $os(x + p) = os(x)$ for $m \leq x < 2m + p$), and that this is "compatible" with the input sequence (that is, when the output sequence advances through one period p in this range, the input sequence advances through a whole number of periods). Then the output sequence is ultimately periodic. (In fact, $os(x + p) = os(x)$ for all $x \geq m$).

With this result, we can establish positive instances of the problem by calculation. Two special cases have been considered. Again, I'm grateful to Neil Calkin for his independent work. The first case concerns inputs which are ultimately 1. We could let F(m) be the maximum output period derived from a sequence whose last zero is in position m, with the convention that $F(m) = \infty$ if a non-periodic sequence occurs. I have calculated F(m) for $m \leq 15$. It begins modestly, but its rate of growth seems to be increasing. (We have F(13) = 260, F(14) = 1479, F(15) = 6016). For m = 16, one of the $2^{15}$ possible inputs has so far defied my attempts to establish its period, despite calculation of more than 60000 terms of output.

The other test case involves purely periodic inputs. Here, an even more dramatic result is found. For input period at most 4, and also for period 5 with three exceptions (those beginning 01001, 01010 and 10010), the output is ultimately periodic with small period. (The largest period is 25, but periods of 25 are made up of five nearly equal blocks of five). However, extensive calculations on the three exceptions have failed to establish a period. (In one case, more than 400,000 terms were computed). Two further observations appear relevant:

(i) The calculation can be used to give lower bounds for the possible period. (If s(n) = 1, then s cannot have ultimate period dividing n). Thus, if the outputs are periodic, the period is much longer than 25. (Lower bounds so far are of order 500).

(ii) The density of the output appears to be very nearly constant. In other words, non-periodicity doesn't seem to be resulting from the sequence "thinning out".

There are various parts of mathematics which suggest analogies for this sort of behaviour: for example, almost periodic functions, Thue sequences, and quasicrystals (Penrose tilings). I haven't been able to establish direct links.

Perhaps the most fascinating analogy is with the concepts of periodicity and chaos in dynamics. There is a difference, in that a dynamical system involves iteration of a measure-preserving transformation on a space. Though we have measure spaces, we have no measure-preserving transformations, and there is no sensible reason for iterating the map computed by the machine $M_{sf}$. But I have hopes that this analogy will be useful. It is especially tantalising that one of the most easily-realised examples of a chaotic dynamical system is the spherical pendulum with periodic driving force (Miles (1984), Tritton (1986)).

There may be still other measures of approximate periodicity. For example, let $g_s(n)$ be the number of different subsequences of length n of the subsequence s, and $g_s^*(n)$ the numbers which occur infinitely often. Thus an ultimately periodic sequence has $g_s$ bounded by a constant, whereas a completely random sequence has $g_s^*(n) = 2^n$ for every n. We could interpret "slow growth of $g_s$ or $g_s^*$" as meaning "s is close to being periodic". What are $g_s$ and $g_s^*$ for our chaotic outputs earlier? What are their possible values for a sf-universal sequence?

## 6 *COMPUTATIONAL ASPECTS*

The simplest practical implementation follows the description of the machine $M_{sf}$, replacing the oracle with the obvious function for checking whether n = x + y with s(x) = s(y) = 1 (for example, a loop or "while" statement is a high-level programming language). There are some simple improvements which have a considerable effect on performance.

In situations involving exhaustive search (such as the computation of $f_{sf}$ or $e_{sf}$, or the "spectroscope"), it is quicker to generate the sum-free prefixes in lexicographic order, going directly from a sequence to its successor. For sequences of fixed length n, this can be done as follows. Maintain both the output sequence s, and a sequence t recording positions x + y for which s(x) = s(y) = 1. To find the successor of a given pair s, t, set i = n. While either s(i) or t(i) is 1, set this value to 0 and decrease i. If we end up at i = 0, the enumeration is finished. Otherwise, set s(i) = 1. Now, we must recompute t. But the values t(j) for $j \leq i$ are unchanged, so we need only work from i + 1 to n; and, if s(x) = 1, then $x \leq i$, so further reductions in the work can be made. Note that, in situations like this,requirements of space are very modest.

By contrast, consider the search for periodicity of a fixed sequence. Here, a large amount of memory is required, and we must reduce this requirement without increasing too much the time taken. Instead of storing the sequence s, record the sizes of the gaps between the positions of successive ones in s. That is, if ones occur in positions $n_1, n_2, n_3, \ldots$, record $n_1, n_2-n_1, n_3-n_2, \ldots$ Empirically, for the output generated by an ultimately periodic input, these numbers do not grow too fast. (In the examples, one byte suffices for each). Also, the representation is computationally efficient. Suppose that $x_1, x_2, \ldots, x_n$ are stored, and the oracle has been asked whether $\Sigma x_i + u = y + z$ where s(y) = s(z) = 1. Set v: = u, i: = 1, j: = n, and repeat the following operation:

if $v \geq 0$ then v: = v - $x_i$, i: = i + 1;
if $v < 0$ then v: = v + $x_j$, j: = j - 1.

The loop terminates if either v = 0 or $j < i - 1$. In the first case, the oracle answers "yes"; in the second case, "no".

The calculations reported on this paper were run on a Sinclair ZX Spectrum, programmed in Pascal or assembly language, using programming tools by Hisoft.

## 7 *CONCLUSION*

I collect here some problems, most of which have been mentioned in the course of the paper.

1. Obtain more information about the Main Conjecture. This requires new insights, though a few more cases of part (i) could probably be settled by more computation.

2. Failing a proof of the Main Conjecture, find out more about density; in particular, establish that the spectrum does consist of discrete "lines", and find the mean and variance.

3. Decide whether ultimately periodic inputs necessarily give ultimately periodic outputs. If not, in what theoretical sense (if any) are the outputs close to periodic?

4. Find the Hausdorff dimension of $\mathcal{Sf}$ and its ½-dimensional measure, and find the asymptotic behaviour of $f_{\mathcal{Sf}}(n)$ and $e_{\mathcal{Sf}}(n)$.

5. Is it true that any sf-universal sequence has density zero?

6. Investigate the functions g and $g^*$ of the last chapter.

7. Perform similar investigations for other classes. Examples include

(i) {S: S contains no n-term arithmetic progression};

(ii) {S: S contains no sequence of n numbers together with all sums of non-empty subsequences};

(iii) {S: $\Gamma$(S) contains no complete graph of size n + 1}.

(The condition in (iii) is that S contains no sequence of n numbers together with all sums of non-empty consecutive subsequences).

8. For general sets X, investigate relations between Hausdorff dimension of X, behaviour of $f_X$ and $e_X$, and recursion-theoretic aspects of $M_X$. In particular, what are the consequences of $M_X$ having a finite-state oracle?

9. Is there any sensible interpretation for the process of iterating the map defined by $M_X$?

10. A more general set-up replaces $2^*$ by the set of vertices of a rooted tree of height $\omega$, and $2^{\mathbb{N}}$ by the set of 1-way infinite paths in the tree (starting at the root). There is one important context for this set-up. Given a countable relational structure X, the tree is the <u>age</u> of X in the sense of Fraïssé (1986), that is, its vertices are the labelled n-element structures which are embedded in X; any path in the tree corresponds to a structure Y <u>younger</u> than X (that is, of smaller age). The category-theoretic concepts arise in Robinson's <u>finite forcing</u>, but the measure-theoretic analogues are largely unexplored. See Bankston & Ruitenberg (1987), Cameron (1987).

In conclusion, I am grateful to a large number of people, either for helpful comments or for sharing my enthusiasm; especially Wilfrid Hodges, for translating between my language and that of the model theorists; Ken Regan, who suggested Baire category to me; Neil Calkin and Dominic Welsh, for performing experiments; the model theorists at Queen Mary College, for remaining cheerful and constructive through a course of lectures on the subject; the dynamical theorists at Queen Mary College, for encouraging me to think that the analogy in Chapter 6 might not be too far-fetched; and Vanessa Tyrrell, for producing the manuscript.

<u>Late Note</u>. The first part of the Main Conjecture has been proved. If T is a complete sum-free set mod n, with $|T| = k$, then I can show that

$$p(E(n,T)) \geq (c/2)^{n-k}$$

where $c = 0.218$ is the probability of *Odd* (Theorem 4.1). The proof, which is not difficult, uses the FKG inequality.

*REFERENCES*

Bankston, P. & Ruitenberg, W. (1987). Notions of relative ubiquity sets of relational structures. To appear.

Cameron P.J. (1984). Aspects of the random graph. In Graph Theory and Combinatorics, ed. B. Bollabás, pp. 65-79. London: Academic Press.

Cameron, P.J. (1985 a). Cyclic automorphisms of a countable graph and random sum-free sets. Graphs Comb. 1, 129-135.

Cameron, P.J. (1985 b). Four lectures on projective geometry. In Finite geometries, ed. C.A. Baker & L.M. Batten, pp. 27-63. New York: Marcel Dekker.

Cameron, P.J. (1987). Ubiquitous relational structures. Proc. conf., Lyon. To appear.

Erdős, P. & Rényi, A. (1963). Asymmetric graphs. Acta Math. Acad. Sci. Hungar., 14, 295-315.

Erdős, P. & Turán. P. (1936). On some sequences of integers. J. London Math. Soc., 11, 261-264.

Falconer, K.J. (1985). The Geometry of Fractal Sets. Cambridge: Cambridge University Press.

Fraissé, R. (1986). Theory of Relations. Amsterdam: North Holland.

Furstenburg, H. (1977). Ergodic behaviour of diagonal measures and a theorem of Szemerédi on arithmetic progressions. J. d'Anal. Math., 31, 204-256.

Gardiner, A.D. (1976). Homogeneous graphs. J. Combinatorial Theory (B), 20, 94-102.

Hanson, D. (1976). Sum-free sets and Ramsey numbers. Discr. Math. 14, 57-61

Hanson, D., & Seyffarth, K. (1984). K-saturated graphs of prescribed minimal degree. Congr. Num., 42, 169-182.

Henson, C.W. (1971). A family of countable homogeneous graphs. Pacific J. Math., 38, 69-83.

Lachlan, A.H. & Woodrow, R.E. (1980). Countable homogeneous undirected graphs. Trans. Amer. Math. Soc., 262, 51-94.

Miles, J. (1984). Resonant motion of a spherical pendulum. Phys. D., 11, 309-323.

Oxtoby, J.C. (1980). Measure and Category. (Second edition). Berlin: Springer-Verlag.

Petersen, K. (1983). Ergodic Theory. Cambridge: Cambridge University Press.

Rado, R. (1964). Universal graphs and universal functions. Acta. Arith., 9, 331-340.

Rogers, C.A. (1970). Hausdorff Measures. Cambridge: Cambridge University Press.

Roth, K.F. (1952). Sur quelques ensembles d'entiers. C.R. Acad. Sci. Paris Sér., A-B, 234, 388-390.

Schur, I. (1916). Über die Kongruenz $x^m + y^m = z^m$. (Mod p). Jber. Deutsch. Math. - Verein., 25, 114-116.

Szemerédi, E. (1969). On sets of integers containing no four elements in arithmetic progression. Acta. Math. Acad. Sci. Hungar., 20, 89-104.

Szemerédi, E. (1975). On sets of integers containing no k elements in arithmetic progression. Acta. Arith., 27, 199-245.

Tritton, D. (1986). Chaos in the swing of a pendulum. New Scientist, 111, 37-40.

van der Waerden, B.L. (1927). Beweis einer Baudet' schen vermutung. Nieuw Arch. Wisk., 15, 212-216.

Wallis, W.D., Street, A.P., & Wallis, J.S. (1972). Combinatorics: Room Squares, Sum-Free Sets, Hadamard Matrices. Berlin: Springer-Verlag.

# PERFECT GRAPHS

V. Chvátal
Department of Computer Science, Rutgers University,
New Brunswick, NJ 08903 USA

In 1960, Claude Berge proposed to call a graph <u>perfect</u> if, for each of its induced subgraphs F, the chromatic number $\chi(F)$ of F equals the clique number $\omega(F)$ of F. At the same time, he conjectured that

(1) a graph is perfect if and only if none of its induced subgraphs is a chordless cycle whose number of vertices is odd and at least five, or the complement of such a cycle.

To stimulate interest in this subject, he also publicized the conjecture that

(2) a graph is perfect if and only if its complement is perfect.

Since (1) implies (2), the two conjectures become known as the Strong and the Weak Perfect Graph Conjectures. The latter conjecture was proved by Lovász (1972a) and is now referred to as the Perfect Graph Theorem; the Strong Perfect Graph Conjecture remains open.

We propose to call a graph <u>Berge</u> if none of its induced subgraphs is a chordless cycle whose number of vertices is odd and at least five, or the complement of such a cycle. In this terminology, (1) states that a graph is perfect if and only if it is Berge; note that the "only if" part holds true (every perfect graph is Berge). Since the class of Berge graphs belongs to coNP by its very definition, (1) implies that

(3) the class of imperfect graphs belongs to NP.

Shortly after proving (2), Lovász (1972b) proved a theorem that implies both (2) and (3):

(4) every minimal imperfect graph F has precisely $\alpha(F)\omega(F)+1$ vertices.

(Here, as usual, $\alpha(F)=\omega(\overline{F})$ with $\overline{F}$ standing for the complement of F.) To see that (4) implies (2), observe that every perfect F has at most $\alpha(F)\omega(F)$ vertices (F can be colored by $\omega(F)$ colors and each color can be assigned to at most $\alpha(F)$ vertices) and that $\alpha(\overline{F})\omega(\overline{F})=\alpha(F)\omega(F)$. To show that (4) implies (3), let us follow Bland, et al. (1979) in calling F *partitionable* if there are integers r and s such that $r\geq 2, s\geq 2$, and for every vertex of F, the vertices of F-v can be partitioned into r cliques of size s as well as into s stable sets of size r. Now (2) implies instantly that every minimal imperfect graph is partitionable; Bland et al. observed that every partitionable F is imperfect. (Since $\omega(F-v)=s$ for all v, any clique of size s+1 in F would have to include all v, which is impossible since $\alpha(F)\geq r\geq 2$. Hence $\omega(F)=s$ and similarly $\alpha(F)=r$; now note that F has rs+1 vertices.) To put it differently, G is imperfect if and only some induced subgraph of G is partitionable. As pointed out by Cameron & Edmonds (1982), this observation implies (4) since the class of partitionable graphs belongs to NP. The other obvious conjecture,

(5) the class of perfect graphs belongs to NP,

remains open, and so does its companion conjecture,

(6) the class of Berge graphs belongs to NP.

(Note that (5) and (6) are different conjectures if and only if (1) is false.)

If (5) holds true then every perfect G admits a succint certificate of perfection; by (2), every certificate of perfection of G certifies perfection of $\overline{G}$ as well. A truly elegant certificate of perfection would make it obvious by its very form that it applies to G if and only if it applies to $\overline{G}$. What properties shared by G and $\overline{G}$ determine perfection? This question led Chvátal (1984b) to the following definition: graphs with the same set of vertices are said to have the same *$P_4$-structure* if each set of four vertices induces a $P_4$ (the chordless path with vertices, a,b,c,d and edges ab,bc,cd) either in both graphs or in neither of them. For example, the two graphs in Fig. 1 have the same $P_4$-structure:

Fig. 1

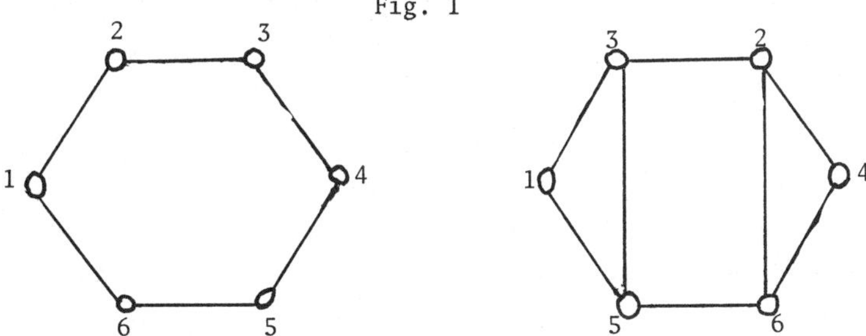


Brute force shows that every graph with the $P_4$-structure of a Berge graph is Berge, and so (1) implies that

(1½) every graph with the $P_4$-structure of a perfect graph is perfect;

since $P_4$ is self-complementary, (1½) implies (2). For this reason, (1½) was known as a Semi-Strong Perfect Graph Conjecture until Reed (1986) proved it in 1984.

(1½) is one guide in the quest for an affirmative answer to (5): a certificate of perfection may refer only to the $P_4$-structure of a graph, without any reference to its other properties. Another guide is the hope that every perfect graph could be decomposed into some "primitive" graphs whose relatively transparent structure would make certifying perfection easy. One result guided by both of these two thoughts involves calling vertices x and y <u>partners</u> if there are vertices a,b,c such that both {x,a,b,c} and {y,a,b,c} induce a $P_4$. The result (Chvátal (1986)) goes as follows:

(7) Let the vertices of G be colored red and white in such a way that every two partners have the same color. Then G is perfect if and only if each of its two subgraphs induced by all the vertices of one color is perfect.

(Testing whether G admits a coloring in which every two partners have the same color <u>and</u> either color appears at least once is easy: first construct the "partner graph" in which two vertices are adjacent iff they are partners, and then test the partner graph for connectivity.)

The proof of (7) relies in part on an earlier result of Chvátal & Hoàng (1985), in which the hypothesis "every two partners have the same color" was replaced by the stronger hypothesis "every $P_4$ has an even number of red vertices". Another formulation of the stronger hypothesis is "there is no $P_4$ of type RWWW or WRWW or WRRR or RWRR", where RWWW stands for the $P_4$ with a red and b,c,d,white, etc.; this formulation suggests the following idea. There are (up to the $P_4$'s symmetry) ten ways of coloring the vertices of a $P_4$ red and white, and so $2^{10}$ sets can be made out of these ten elements. For each of these sets S, consider the analogue of (7) in which the hypothesis "every two partners have the same color" is replaced by the hypothesis "there is no $P_4$ of a type that belongs to S". Some of these 1,024 statements will be false; among the true ones, some will be weaker than others. Chvátal et al. (1986) proved that there are precisely twelve strongest theorems of this form: six of them arise by setting

S={RWWW, WRWW, WRRR, RWRR},
S={RRRR, WRRW, RWWR, WWWW},
S={WRRR, WRRW, RWWR, WWWW, RWRR},
S={WRRR, WRRW, RWWR, WWWW, RWRW},
S={WRRR, WRRW, RWWR, RWWW, RWRW},
S={WRRR, WRRW, WRWW},

and the remaining six arise from these six by substituting $\overline{G}$ for G or by switching colors. (Note that the first of these twelve theorems is nothing but the Chvátal-Hoàng theorem, and that all the others require more information about G than just its $P_4$-structure.)

A different line of attack on the Strong Perfect Graph Conjecture consists of finding more and more properties of minimal imperfect graphs in the hope that eventually the only graphs with all these properties will be chordless cycles whose number of vertices is odd and at least five, and the complements of these cycles. As we remarked earlier, every minimal imperfect graph is partitionable; what else do we know about minimal imperfect graphs? Padberg (1974) proved that, with n standing for the number of vertices in a minimal imperfect graph G,

(8) all the cliques of size $\omega(G)$ in G can be enumerated as $C_1,C_2,\ldots,C_n$ and all the stable sets of size $\alpha(G)$ can be enumerated as $S_1,S_2,\ldots,S_n$ in such a way that $C_i \cap S_j = \emptyset$ if and only if $i=j$.

Chvátal (1985) proved that

(9) there is no set T of vertices such that G-T is disconnected and some vertex in T is adjacent to all the remaining vertices in T.

(A slightly weaker theorem was proved earlier by Tucker (1977) and, with $\overline{G}$ in place of G, by Olaru (1969).) Meyniel (1986) proved that

(10) between every two vertices there is a chordless path with an odd number of edges.

However, none of these three properties restrict the list of candidates for minimal imperfect graphs any further than the property of being partitionable: Bland et al. proved that (8) holds whenever G is partitionable, Chvátal's proof shows that (9) holds whenever G is partitionable, and Reed (personal communication) proved that (10) holds whenever G is partitionable.

Which properties then distinguish minimal imperfect graphs from those that are merely partitionable? Chvátal et. al. (1984) pointed out that, in a minimal imperfect graph,

(11) no set of fewer than $\alpha(G)+\omega(G)$ vertices meets all cliques of size $\omega(G)$ and all stable sets of size $\alpha(G)$.

Chvátal & Sbihi (1986) proved that

(12) there are no disjoint sets S,T of vertices such that each vertex outside $S \cup T$ is adjacent either to all the vertices in S or to none of them, each vertex outside $S \cup T$ is adjacent either to all the vertices in T or to none of them, and $\max(|S|,|T|) \geq 2$, $|S|+|T| \leq n-2$.

(Their proof of (12) relies in part on an earlier result of Olariu (1986b) who required $|S|+|T|=n-2$.) Chvátal (1985) conjectured that

(13) the set of vertices cannot be partitioned into nonempty S and T so that S induces a disconnected subgraph in G and T induces a disconnected subgraph in $\overline{G}$;

this conjecture remains open. The partitionable graph shown in Fig. 2 has none of these three properties: (11) is contradicted by {0,2,4,6,8}, (12) is contradicted by S={1,2,8,9},T={3,4,6,7}, and (13) is contradicted by S={9,0,1,4,5,6},T={2,3,7,8}.

Fig. 2

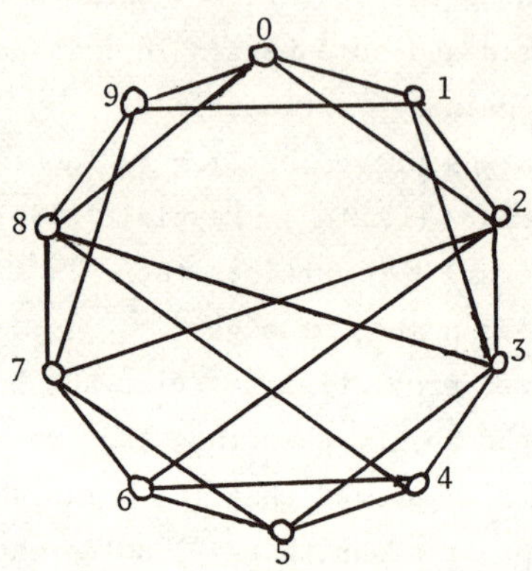


Finally, evidence in support of the Strong Perfect Graph Conjecture is provided by theorems asserting that Berge graphs are perfect as long as they satisfy some additional requirements. For instance, Berge graphs containing no induced subgraph F are known to be perfect when F is the claw (Parthasarathy & Ravindra (1976)), the tetrahedron (Tucker (1977)), the diamond (Tucker (1986)), and the bull (Chvátal & Sbihi (1986)).

Fig. 3

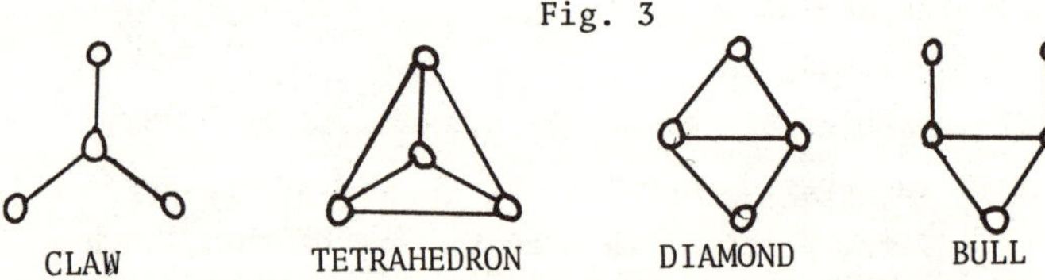


One may also forbid more than one induced subgraph at a time: Berge graphs are known to be perfect if they contain neither the chordless path with six vertices nor its complement (Hayward (1986)), if they contain no odd cycle with precisely one chord (Meyniel (1976)), if they contain neither chordless cycles with at least six vertices nor their complements (Hayward (1985)), and if they are planar (Tucker (1973)). A different kind of these theorems involves classes of Berge graphs that are defined by their membership certificates. Three examples are the *perfectly orderable graphs* (whose vertex-sets admit a linear order $<$ such that no induced $P_4$ has $a<b, d<c$), the *opposition graphs* (whose vertex-sets admit a linear order $<$ such that no induced $P_4$ has $a<b, c<d$), and *alternately orientable graphs* (whose edges admit directions that alternate on every chordless cycle with at least four vertices). These graphs have been proved perfect in (Chvátal (1984a)), (Olariu (1986a)), and (Hoàng (1986)), respectively.

REFERENCES

Bland, R.G., Huang, H.-C. & Trotter, L.E., Jr. (1979). Graphical properties related to minimal imperfection. Discrete Math., 27, 11-22.

Cameron, K.G. (1982). Polyhedral and Algorithmic Ramifications of Antichains. Ph.D. Thesis, University of Waterloo.

Chvátal, V. (1984a). Perfectly ordered graphs. In Topics on Perfect Graphs, eds. C. Berge & V. Chvátal. Annals of Discrete Math., 21, 63-5.

Chvátal, V. (1984b). A semi-strong perfect graph conjecture. In Topics on Perfect Graphs, eds. C. Berge & V. Chvátal. Annals of Discrete Math., 21, 279-80.

Chvátal, V. (1985). Star-cutsets and perfect graphs. J. Combinatorial Th. (B), 39, 189-99.

Chvátal, V. (1986). On the $P_4$-structure of perfect graphs. III. Partner decompositions. J. Combinatorial Th. (B), to appear.

Chvátal, V., Graham, R.L., Perold, A.F. & Whitesides, S.H. (1984). Combinatorial designs related to the Strong Perfect Graph Conjecture. In Topics on Perfect Graphs, eds. C. Berge & V. Chvátal. Annals of Discrete Math., 21, 197-206.

Chvátal, V. & Hoàng, C.T. (1985). On the $P_4$-structure of perfect graphs. I. Even decompositions. J. Combinatorial Th. (B), 39, 209-19.

Chvátal, V., Lenhart, W.J. & Sbihi, N. (1986). Two-colorings that decompose perfect graphs. J. Combinatorial Th. (B), submitted.

Chvátal, V. & Sbihi, N. (1986). Bull-free Berge graphs are perfect. Graphs and Combinatorics, to appear.

Hayward, R.B. (1985). Weakly triangulated graphs. J. Combinatorial Th. (B), 39, 200-8.

Hayward, R.B. (1986). Murky graphs are perfect. J. Combinatorial Th. (B), submitted.

Hoàng, C.T. (1986). Alternately orientable graphs. J. Combinatorial Th. (B), to appear.

Lovász, L. (1972a). Normal hypergraphs and the perfect graph conjecture. Discrete Math., 2, 253-67.

Lovász, L. (1972b). A characterization of perfect graphs. J. Combinatorial Th. (B), 13, 95-8.

Meyniel, H. (1976). On the perfect graph conjecture. Discrete Math., 16, 339-42.

Meyniel, G. (1986). A new property of critically imperfect graphs and some consequences. European J. Combinatorics, to appear.

Olariu, S. (1986a). All variations on perfectly ordered graphs. J. Combinatorial Th. (B), to appear.

Olariu, S. (1986b). No antitwins in minimal imperfect graphs. J. Combinatorial Th. (B), submitted.

Olaru, E. (1969). Über die Überdeckung von Graphen mit Cliquen. Wiss. Tech. Hochsch. Ilmenau. See also E. Olaru & H Sachs, Contributions to a characterization of the structure of perfect graphs. In Topics on Perfect Graphs, eds. C. Berge & V. Chvátal. Annals of Discrete Math., 21, 121-44 (1984).

Padberg, M.W. (1974). Perfect zero-one matrices. Math. Programming 6, 180-96.

Parthasarathy, K.R. & Ravindra, G. (1976). The strong perfect graph conjecture is true for $K_{1,3}$-free graphs. J. Combinatorial Th. (B), 21, 212-23.

Reed, B. (1986). A semi-strong perfect graph theorem. J. Combinatorial Th. (B), to appear.

Tucker, A.C. (1973). The strong perfect graph conjecture for planar graphs. Canad. J. Math. 25, 103-14.

Tucker, A.C. (1977). Critical perfect graphs and perfect 3-chromatic graphs. J. Combinatorial Th. (B), 23, 143-49.

Tucker, A.C. (1986). Coloring $(K_4$-e)-free graphs. J. Combinatorial Th. (B), to appear.

# MY JOINT WORK WITH RICHARD RADO

Paul Erdös
Mathematical Institute of the Hungarian Academy of Sciences
Budapest, Hungary

and

AT&T Bell Laboratories
Murray Hill, New Jersey 07974

I first became aware of Richard Rado's existence in 1933 when his important paper *Studien zur Kombinatorik* appeared. I thought a great deal about the many fascinating and deep unsolved problems stated in this paper but I never succeeded to obtain any significant results here and since I have to report here about our joint work I will mostly ignore these questions. Our joint work extends to more than 50 years; we wrote 18 joint papers, several of them jointly with A. Hajnal, three with E. Milner, one with F. Galvin, one with Chao Ko, and we have a book on partition calculus with A. Hajnal and A. Máté. Our most important work is undoubtedly in set theory and, in particular, the creation of the partition calculus. The term partition calculus is, of course, due to Rado. Without him, I often would have been content in stating only special cases. We started this work in earnest in 1950 when I was at University College and Richard in King's College. We completed a fairly systematic study of this subject in 1956, but soon after this we started to collaborate with A. Hajnal, and by 1965 we published our GTP (Giant Triple Paper - this terminology was invented by Hajnal) which, I hope, will outlive the authors by a long time. I would like to write by centuries if the reader does not consider this as too immodest. Since this conference is more on finite combinatorics, I will speak more on our work on finite sets and where possible, I will restrict myself to countable sets. I have another reason for this: very many new results were proved in this subject, much of it using mathematical logic and forcing. Many mathematicians (e.g. Hajnal, Shelah, Galvin, Laver, Baumgartner and many others) are more competent than I to write about these results. Hajnal and I [1] published two long survey papers about 15 to 20 years ago about many of our solved and unsolved problems in set theory. Clearly a new paper or papers, perhaps a book on this subject, would be very desirable, but as I just stated, many others are now more competent than I to write such a paper or book. I will list in the references our joint papers with Richard and will refer to them by their number.

I started to correspond with Richard in late 1933 or early 1934 when he was a German refugee in Cambridge. We first met on October 1, 1934 when I first arrived at Cambridge from Budapest. Davenport and Richard met me at the railroad station in Cambridge and we immediately went to Trinity College and had our first long mathematical discussion.

In one of my first letters to Richard early in 1934, I posed the following question: Let $S$ be an infinite set of power $m$. Split the countable subsets of $S$ into two classes. Is it true that there always exists an infinite subset $S_1$ of $S$ all of whose countable subsets are in the same class? This, if true, would be a far reaching generalization of Ramsey's theorem. Almost by return mail, Rado found the now well-known counterexample using the axiom of choice. Later on all this led to many interesting developments. The answer to my question becomes affirmative if we restrict the partition in various ways, e.g., we only permit Borel or analytic partitions, or if we do not permit the use of the axiom of choice. These results are connected with the names of Galvin, Prikry, Silver, Mathias and others. Magidor and I much later used some of these results. [2]

Actually our first joint paper was done with Chao Ko and was essentially finished in 1938. Curiously enough it was published only in 1961. One of the reasons for the delay was that at that time there was relatively little interest in combinatorics. Also in 1938, Ko returned to China, I went to Princeton and Rado stayed in England. I think we should have published the paper in 1938. This paper [XI] "Intersection theorems for systems of finite sets" became perhaps our most quoted result. Our principal result states as follows: Let $|S| = n$, $n \geqslant 2k$, $A_i \subset S$, $1 \leqslant i \leqslant t(n;k)$, be an intersecting family of subsets of $S$ i.e., $A_i \cap A_j \neq \varnothing$ for every $1 \leqslant i \leqslant j \leqslant t(n;k)$. If we further assume $|A_i| = k$, then

$$t(n;k) \leqslant \binom{n-1}{k-1} \tag{1}$$

and there is equality in (1) if and only if all our sets $A_i$ have a common element. We in fact proved that (1) holds if $|A_i| = k$ is replaced by $|A_i| \leqslant k$, $A_i \not\subset A_j$, i.e., that our family forms a Sperner system. We also proved that if $n > n_0(k,r)$ and we assume $|A_i \cap A_j| \geqslant r$, $1 \leqslant i \leqslant j \leqslant t(n;k,r)$, then

$$\max t\,(n;k,r) \geqslant \binom{n-r}{k-r} \qquad (2)$$

Min first observed that the assumption $n > n_0\,(k,r)$ is really needed and our crude upper bound for $n_0\,(k,r)$ has been gradually improved by Frankl and Wilson and the exact value is now known. At the end of our paper, several further results are proved and problems are posed, which became very popular. Most of them were settled by Katona, Kleitman and others. Let me state one problem stated in our paper which seems deceptively simple but no progress has been made with it. Let $|S| = 4n$, $|A_i| = 2n$, $A_i \subset S$, $1 \leqslant i \leqslant T(n)$. Assume, $|A_i \cap A_j| \geqslant 2$, $1 \leqslant i \leqslant j \leqslant T(n)$. Is it then true that

$$\max T(n) = \left( \binom{4n}{2n} - \binom{2n}{n}^2 \right) /2 \ ? \qquad (3)$$

It is easy to see (and is contained in our paper) that (3) if true is best possible. This uses the original idea of Min. Let the set $S$ be the integers $1, 2, \ldots, 4n$. $S_1 \cup S_2 = S$, $|S_1| = |S_2| = 2n$. The $A_i$'s satisfy $|A_i| = 2n$, $|A_i \cap S_1| \geqslant n + 1$.

I offer 250 pounds for the proof or disproof of (3). This problem is more than 45 years old.

Due to our rapidly advancing age, it is doubtful if all three of us will ever be together again, (at least not in this world). I had the good fortune to see Ko in June 1986 in Beijing and found him in good health. I, of course, see Richard often and speak to him on the phone every few months.

As far as I know, Hilton and Milner wrote the first paper on our theorem. They proved that if $|S| = n$, $n \geqslant 2k$, $|A_i| = k$, is an intersecting family which is not a clique (i.e. there is no element which is contained in all the $A$'s) then the size of our family is at most

$$\binom{n-1}{k-1} - \binom{n-k-1}{k-1} + 1 \qquad (4)$$

and (4) is best possible. If we assume that our family $A$ is intersecting but there is no set of $r$ points so that one of the $A$'s contains one of these points then for $r \geqslant 2$ the maximum size of our family is not known. Perhaps $r = k - 1$ is the most interesting case. See our paper with Lovász [3] and many papers of P. Frankl, Kleitman, Katona, Füredi, Pyber and others and also, a forthcoming paper of Aigner, Andreae and myself.

To end the discussions of this problem, I would like to state a few problems. In our paper we notice that if $|S| = n$, and $F$ is an intersecting family of subsets of $|S|$ then trivially $\max F = 2^{n-1}$. We noticed that there are many such families which are not cliques. Hindman and I obtained upper and lower bounds for the number of such families ([4]).

We further observed that if $|F| = 2^{n-1}$ and every three sets in $F$ have a non-empty intersection then $F$ must be a clique.

Frankl and Füredi now considered the following problem: Let $|S| = n$, $F$ a family of subsets and assume that every $k$ of them have an intersection of size $\geqslant r$. Determine max $F$. In particular for which values of $k$ and $r$ is it true that max

$F = 2^{n-r}$? Frankl proved that this holds for $k = 2$, $r = 2$ and he conjectured that this holds also for $k = 3$, $r = 2$. Füredi pointed out to me that this certainly does not hold for $k = 2$ and $r$ sufficiently large. Clearly many unsolved problems remain.

Here is a nice conjecture of Frankl and Pach: Let $|S| = n$, $|A_i| = k$, $1 \leqslant i \leqslant t_n$, be an intersecting family. If $t_n > \binom{n-1}{k-1}$ then there is an $A_j$ so that for every $B \subset A_j$ there is an $A_i$ for which $A_i \cap A_j = B$. They can prove this if $t_n > \binom{n}{k-1}$.

The final problem which I believe was first raised by Rothschild, Szemerédi and myself is this. Let $|S| = n$, and let $F$ be an intersecting family of sets with $|A_i| = k$. Assume that every point of $S$ has degree not exceeding $cF$, $c < 1$. What can one say about max $|F|$? The point of our degree condition is that no element can be contained in too many of our sets, i.e., our family $F$ is very far from being a clique. Füredi solved this problem for many values of $c$ ([5]).

Our first joint paper which actually appeared is on the canonical Ramsey theorem [I] which was the first paper in this direction and which also had a great deal of influence (see, e.g., the work of Graham, Leeb and Rothschild, the work of Deuber, Voigt, Prömel and Laufmann; and the book of Graham, Rothschild and Spencer on Ramsey theory [6]).

Here is our Theorem: Let $N$ be the set of integers. Split the $r$-tuples of the integers into any number of classes. Then there always is a $k$, $1 \leqslant k \leqslant r$, and an infinite subset $u_1 < u_2 < \ldots$ of $N$ so that our distribution is canonical on $u_1 < u_2 < \ldots.$ In other words, two $r$-tuples $u_{i_1} < u_{i_2} < \ldots < u_{i_k}; v_{j_1} < v_{j_2} < \ldots < v_{j_k}$ are in the same class if and only if $u_{i_{v_1}} = v_{j_{v_1}}, \ldots, u_{i_{v_k}} = v_{j_{v_k}}$. There are clearly $2^r$ canonical distributions and we obtain the classical Ramsey's theorem if the number of classes in finite; then there is only one canonical distribution when $k = n$, i.e., all $r$-tuples belong to the same class. Rado and I extended our theorem for infinite cardinals and Galvin and Taylor improved our results and, in fact, obtained best possible versions of it. Many generalizations and extensions of our theorem with Richard were obtained by many mathematicians, e.g., Deuber, Voigt, Laufmann and many others.

Here I state the canonical van der Waerden theorem which was found independently by R. L. Graham and myself: Split the integers into a finite or infinite number of classes. Then for every $k$ there is an arithmetic progression of length $k$ all of whose terms are in the same class or all of whose terms are in different classes. I was told of this problem by Dr. Wilkie after a lecture of mine at the Open University. Both Graham and I used in our proof Szemerédi's celebrated theorem: every sequence of positive density contains arbitrarily long arithmetic progressions. Nešetřil and Rödl later found a proof of the canonical van der Waerden theorem which does not use Szemerédi's theorem.

Perhaps our canonical theorem and also later the discovery of the partition calculus explains the success of our collaboration. I was good at discovering perhaps difficult and interesting special cases and Richard was good at generalizing them and putting them in their proper prospective.

Now let me state one of our minor results [II]: A sequence $a_1 < a_2 <\dots$ has property $S$ if every infinite subsequence has two terms, one of which divides the other. We proved that if $a_1 < a_2 <\dots$ has property $S$ then the set of integers $\Pi a_i^{\alpha_i}$, $0 \leqslant \alpha_i < \infty$ also has property $S$. It turned out that Higman already had a more general result. In [VIII] we proved some results on partially well ordered sets of vectors and later Nash-Williams and others proved much more general and important results.

But now I have to return to serious Mathematics. Perhaps our most fascinating unsolved problem in finite combinatorics is about $\Delta$-systems [IX]. A family of sets $\{A_\alpha\}$ is called a $\Delta$-system if the intersection of any two of the $A$'s equals $\cap A_\alpha$, i.e., the intersection of any two of them equals the intersection of all of them. Now we investigated and completely solved in the infinite case the following problem: Let $\{A_\alpha\}$ be a family of $m$ sets each of size $\leqslant n$. When must it contain a $\Delta$-system of size $p$? Our results had applications in topology, set theory and logic (e.g., in forcing). We solved the infinite problems even without the use of the continuum hypothesis. But if $p$, $m$ and $n$ are finite, surprising difficulties arise. Denote by $f(n;p)$ the smallest integer such that every family of $f(n;p)$ sets of size $n$ contains a subfamily of size $p$ which forms a $\Delta$-system. Even for $p = 3$ the problem seems to be very difficult. I offer 1000 dollars for the proof or disproof of our old conjecture:

$$f(n;3) < c^n . \qquad (5)$$

We only proved $f(n;3) < 2^n n!$ and $f(n;p) < (p-1)^n n!$ This was improved by Spencer to $f(n;3) < (1+o(1))^n n!$. $f(n;3) > 2^n$ was proved in [IX] and is in fact very easy. Let $x_i$, $y_i$, $1 \leqslant i \leqslant n$ be $2n$ elements. Our $2^n$ sets are defined as follows: Each of them contains exactly one of the elements $x_i, y_i$; $1 \leqslant i \leqslant n$. Clearly these $2^n$ sets do not contain a Δ-system of 3 members. Abbott and Hanson improved our bound by showing $f(n;3) > 10^{n/2}$, and Abbott showed $f(3,3) = 21$. I would like to call attention to the fact that it is not yet known that for $n > n_0$,

$$f(n;3) < n! \tag{6}$$

A family of sets $\{A_\alpha\}$ is called a weak Δ-system if the intersection of any two of our sets has the same size. In a triple paper with Milner [XVI] using the continuum hypothesis we solved all the infinite problems with the help of Hajnal, but

$$f_w(n;3) < c^n \tag{7}$$

is still open, where $f_w(n;3)$ is the smallest integer $u$ so that every family of $u$ sets of size $n$ contains three sets which form a weak Δ-system. As far as I know even $f_w(n;3) < n!$ has not yet been proved and I am sure that $f_w(n;3) < (n!)^{1-\epsilon}$ is still open. After we wrote our paper we found out that Sanjin in the 1940's proved that if $m$ is a regular cardinal and $F$ is a family of $m$ finite sets then this family contains a Δ-system of size $m$. Sanjin used this result in set-theoretic topology. The inequality

$$f(n;k) < (k-1)^n n!$$

is the basis of the star method which was developed by Frankl and Füredi for solving extremal problems concerning several intersecting type families. It will be surveyed in a forthcoming book by Frankl, Füredi and Katona. For a generalization of Δ-systems see the paper of Frankl and Pach [7]. As far as I know $f_w(n;3)^{1/n} \to 2$ has not yet been disproved.

In another paper [III] where we already started to discuss problems of partition calculus we obtained the first reasonable upper bounds for the general Ramsey function. Denote by $f(n;k,r)$ the smallest number so that if we divide the $r$-tuples of the integers $1 \leqslant j \leqslant f(n;k,r)$ into $k$ classes then there always is a subset of the integers $\leqslant f(n;k,r)$ of size $n$ all whose $r$-tuples are in the same class. Before our paper there were only very poor upper bounds for $f(n;k,r)$ for $r > 2$. By using a ramification system we obtained an upper bound for $f(n;k,r)$ as an $(r-1)$-times iterated exponential. In fact we proved

$$f(n;k;r)^{1/n} < k^{k^{\cdot^{\cdot^{k}}}}\Big]\, r-1\,.$$

Hajnal, Rado and I [XII] later showed that in fact $f(n;k,r)$ is greater than an $(r-2)$-times iterated exponential. It seems likely that the $(r-1)$-times iterated exponential gives the correct bound. Let us restrict ourselves for the moment to $r = 3$. The probability method gives without any difficulty

$$f(n;2,3) > 2^{cn^2} \tag{8}$$

and Hajnal proved more than 20 years ago that (see [XVIII])

$$f(n;4,3) > 2^{c \cdot 2^{n/2}} \tag{9}$$

Hajnal and I have a slightly better upper bound then (8) for $f(n;3,3)$. Probably

$$f(n;2,2) > 2^{2^{cn}} \tag{10}$$

but, unfortunately, on this no progress has been made.

This is one of the outstanding open problems of the subject. I offer 500 pounds for a proof or disproof of (10). If (10) holds then it would follow from our methods that $f(n;k,r)$ increases as an $(r-1)$-times iterated exponential.

Hajnal and I have a forthcoming new paper on $f(n;k,3)$ which will show that in many ways $f(n;k,3)$ behaves differently then $f(n;k,2)$ which perhaps will further increase the interest in the fundamental conjecture (10).

In [III] we give the first non-trivial lower bound for the van der Waerden function $W(n)$. $W(n)$ is the smallest integer for which if we divide the integers not exceeding $W(n)$ into two classes then at least one of the classes contains an arithmetic progression of $n$ terms. The only known upper bound for $W(n)$ increases as fast as Ackermann's function. We easily showed by the probability method that

$$W(n) > 2^{n/2}$$

This was improved by Wolfgang Schmidt to $W(n) > 2^{(1+o(1))n}$. Berlekamp showed that if $n = p$ is a prime then $W(p+1) > p2^p$. Lovász and I noticed [3] that the local Lemma of Lovász gives, for every $n$, $W(n) > c2^n/n$. In some of my papers I somewhat carelessly stated that in fact we get $W(n) > c2^n$. As far as I know it has never been proved (see e.g., the survey paper of Graham and Rödl which appears in the same volume as this paper.) The first task would be to prove

$$W(n) > c2^n \tag{11}$$

and then to prove

$$W(n)/2^n \to \infty \tag{12}$$

(11) and (12) will perhaps not be difficult and I offer 25 pounds for a proof. It seems very likely that in fact $W(n)^{1/n} \to \infty$, but perhaps the proof will require a significant new idea.

For a long time all of us believed that $W(n)$ certainly increases much slower than Ackermann's function. As far as I know Solovay was the first who expressed doubts about this. The large majority still believes that the order of magnitude of $W(n)$ is much less than Ackermann's function. The very surprising results of Paris and Harrington showed that simple combinatorial problems can lead to functions which increase much faster than Ackermann's function. Denote by $f^*(n;k,r)$ the smallest integer for which if we divide the $r$-tuples of the integers not exceeding $f^*(n;k,r)$ into $k$ classes then there always is a sequence $a_1 < a_2 < \ldots < a_n \leqslant f^*(n;k,r)$, $a_1 < n$, all whose $r$-tuples are in the same class. The harmless looking extra condition $a_1 < n$ changes the situation completely. Paris and Harrington first show by a simple compactness argument that $f^*(n;k,r)$ is finite, but then comes the surprise: $f^*(n;k,r)$ increases much faster than Ackermann's function, and in fact the existence of $f^*(n;k,r)$ cannot be proved from the Peano axioms. They proved that for every fixed $k$ and $r$ this is possible, but that no such proof exists for all values of $k$ and $r$. It was of course known since Gödel that this situation is possible but it was a great surprise and a great achievement that such simple problems can lead to such seemingly paradoxial results. Many other results of this type are now known. To end this discussion I just remark that Mills and I proved that

$$2^{2^{\frac{n}{2}(1-\epsilon)}} < f^{*}(n;2,2) < 2^{n!^2} \tag{13}$$

Solovay and Ketonen proved that if $k$ or $r$ are $> 2$ then $f^{*}(n;k,r)$ grows already at least as fast as Ackermann's function.

In [III] we also considered the following interesting problem: Let $|S| = m \geqslant \aleph_0$ be an infinite set. For which $m$ is it possible to divide all finite subsets of $S$ into two classes in such a way that every infinite subset $S_1$ of $S$ contains two finite subsets of $S_1$ of the same size which belong to different classes? We proved that this is possible if $|S| \leqslant c$ (the cardinality of the continuum). Hajnal and I [9] later proved that this is in fact possible if the power of $S$ is less than the first strongly inaccessible cardinal, but that it can not be done if $|S|$ is a measurable cardinal. (In those dark and prehistoric times it was not yet known that the first strongly inaccessible cardinal can not be measurable i.e. it was BHT [before Hanf-Tarski]). Silver later proved that our decomposition is in fact possible for very much larger cardinals. He showed that the first cardinal for which such a decomposition is impossible is much larger than the smallest weakly compact cardinal (i.e., for which $m \to (m,m)_2^2$ holds), but it is much smaller than the first measurable cardinal. These investigations of Silver and others had important applications in mathematical logic.

In our paper [VI] we started a systematic investigation of the partition relation $a \to (b_1, b_2)^2_2$ and its generalizations. The partition symbol first occurs in [IV]. In this discussion we will restrict ourselves as much as possible to denumerable sets but first of all I must give the general definition of the partition symbol (see, e.g., our book [XVIII].)

$$a \to (b_h)^r_{h \in H} \tag{14}$$

where $a$ and $b_h$ is a cardinal or an ordinal or an order type, and $H$ is an arbitrary set. In human language (14) means that if we divide the $r$-tuples of $a$ into $H$ classes in an arbitrary way then for some $h \in H$ there is a subset of type $b_h$ all whose $r$-tuples are in the same class. $a \not\to (b_h)^r_{h \in H}$ means that one can split the $r$-tuples into $H$ classes so that these should be no set of type $b_h$ all whose $r$-tuples are in the same class. Before we started our investigations several results of this kind were already known, but were expressed in a different language.

$$\aleph_0 \to (\aleph_0)^r_k \ (r < \aleph_0, k < \aleph_0)$$

is Ramsey's theorem. In human language: if we split the $r$-tuples of a denumerable set into $k$ classes ($r$ and $k$ are finite) then there always is an infinite set all whose $r$-tuples are in the same class.

$$c \not\to (\aleph_1, \aleph_1)^2_2$$

is a well known result of Sierpinski (which was discovered a little later independently by Kurepa). It states that one can partition the pairs of real numbers into two classes so that every subset of power $\aleph_1$ contains a pair from both classes. Finally if $m$ is any infinite cardinal than Dushnik, Miller and I proved

$$m \to (m, \aleph_0)^2_2$$

Perhaps it is nicer to express this result in the language of graphs. If $G$ is a graph of $m$ vertices which does not contain a complete subgraph of $m$ vertices then it contains an infinite independent set. As stated previously, the partition symbol introduced by Rado proved to be immensely useful in expressing in a clear and short way many old and new results and lead to many new problems.

Hajnal, Rado and I in [XII] obtained many further results on the partition symbol and its generalizations and raised a very large number of new and interesting problems which had a great deal of influence on the development of set theory. Many of the problems posed in our papers were solved positively or negatively by Prikry, Galvin, Laver, Baumgartner, Larson, Milner, Shelah, Todorcevic and many others. In many cases undecidability raised its ugly head and more and more often our problems turned out to be undecidable. I personally regret this but at the moment (and perhaps forever) we have to accept it as a fact.

One of our first results in [IV], [V] and [VI] was $\eta \to (\eta, \aleph_0)^2_2$. In fact we obtained a slightly stronger result. If $G$ is a graph whose vertices are the rational numbers, then either $G$ contains an infinite complete graph or an independent set which is dense in an interval. Later this result was extended and generalized enormously by Galvin and Laver. Our next result was $\lambda \to (\omega+n, \omega+n)^2_2$ where $\lambda$ denotes the order type of the continuum. In 1951-52 I was at University College and Richard at King's College. Davenport arranged that my room at College should have a telephone and we had endless mathematical discussions on the 'phone.

We conjectured that for every ordinal $\alpha < \omega_1, \lambda \rightarrow (\alpha, \alpha)^2_2$ and $\omega_1 \rightarrow (\alpha, \alpha)^2_2$, but could not even prove $\lambda \rightarrow (\omega 2, \omega 2)^2_2$. This was proved by Hajnal and a little later Galvin proved $\lambda \rightarrow (\alpha, \alpha)^2_2$.

We first heard of Martin's axiom in the late 1960's from Juhász and at first we did not take it too seriously. But then Hajnal and Baumgartner proved $\omega_1 \rightarrow (\alpha, \alpha)^2_2$ by Martin's axiom and then observed that if it follows from Martin's axiom it is in fact true in ZFC. This triumph of course changed our opinion on Martin's axiom. Later Galvin proved a result stronger than $\omega_1 \rightarrow (\alpha, \alpha)^2_2$ without using Martin's axiom.

In those early days we thought that perhaps for every $\alpha < \omega_1, \omega^\alpha \rightarrow (\omega^\alpha, n)^2_2$. The truth turned out to be much more complicated. First of all Specker proved that $\omega^2 \rightarrow (\omega^2, n)^2_2$. I hope the reader will forgive a very old man for some reminiscences. I passed through Zurich on the way to Israel in November 1934. I met Specker, already an old friend, at the ETH and told him that I offer 20 dollars for a proof or disproof of our conjecture with Richard $\omega^2 \rightarrow (\omega^2, n)^2_2$. A few days later Specker sent me his now well known proof of the conjecture. At first I thought that I can prove $\omega^n \rightarrow (\omega^n, 3)^2_2$, but in fact the proof only gave $\omega^{2n} \rightarrow (\omega^{n+1}, 4)^2_2$ and soon Specker found his well known counterexample $\omega^n \not\rightarrow (\omega^n, 3)^2_2$ for all $n$, $3 \leqslant n < \omega$. Neither Specker's proof nor his counterexample worked for

$$\omega^\omega \rightarrow (\omega^\omega, 3)_2^2 \qquad (15)$$

In the late 1960's I offered 250 dollars for a proof or disproof of (15). Chang in 1969 proved (15) and I gladly handed him the well deserved prize. Milner somewhat simplified Chang's very complicated proof and also showed $\omega^\omega \rightarrow (\omega^\omega, n)_2^2$. Finally Jean Larson obtained a relatively simple proof of $\omega^\omega \rightarrow (\omega^\omega, n)_2^2$ and proved many related results. The first open problem now is: $\omega^{\omega^2} \rightarrow (\omega^{\omega^2}, n)_2^2$.

I offer 250 pounds for a proof or disproof of $\omega^{\omega^2} \rightarrow (\omega^{\omega^2}, 3)_2^2$ and 1000 pounds for the complete characterization of the values of $\alpha$ for which

$$\omega^{\omega^\alpha} \rightarrow (\omega^{\omega^\alpha}, n)_2^2$$

holds. We conjectured that if $\alpha \rightarrow (\alpha, 3)_2^2$ holds then for every $n$ also $\alpha \rightarrow (\alpha, n)_2^2$ holds. This conjecture if true would be useful both for finite and infinite combinatorics. The following problem should be mentioned here: For which values of $n$, $k$ and $l$ does

$$\omega^n \rightarrow (\omega^k, l)_2^2$$

hold? Galvin, Hajnal and independently Haddad and Sabbagh reduced this problem to a finite combinatorial problem and Eva Nosal nearly completely settled it.

In [XII] if we use the continuum hypothesis and exclude large cardinals and restrict ourselves to cardinal numbers we settled nearly all the problems about the truth of the partition relation $a \to (b,c)_2^2$. In our book [XVIII] we give a fairly detailed investigation how far we can get if the continuum hypothesis is dropped, but as stated earlier I ignore here these investigations. I would only like to mention one striking open problem. In [XII] we prove that if $c = \aleph_1$ then $c \nrightarrow [c]_c^2$. In other words one can color the pairs of a set of power $\aleph_1$ by $\aleph_1$ colors so that every subset of power $\aleph_1$ has edges of all the colors. In fact we show that every bipartite $K(\aleph_1,\aleph_0)$ contains edges of all the colors. Let us now not admit the continuum hypothesis. Is it true that $c \to [\aleph_1]_3^2$ holds? In other words can one color the pairs of real number by three colors so that every set of power $\aleph_1$ should contain edges of all three colors? In view of the simplicity of the old proof of Sierpinski for $c \nrightarrow (\aleph_1,\aleph_1)_2^2$ it is very surprising why this simple question should be so difficult. In fact it is generally believed that the problem is undecidable. Galvin and Shelah proved $\aleph_1 \nrightarrow [\aleph_1]_4^2$ and $2^{\eta_0} \nrightarrow [2^{\eta_0}]_{\eta_0}^2$. Very recently Todorcevic proved $\aleph_1 \nrightarrow [\aleph_1]_{\eta_1}^2$. This certainly is an unexpected and sensational result.

In [VI] we proved

$$\lambda \to (\omega+n,\ 4)_2^3 \qquad (16)$$

The proof was quite complicated and we never could get a stronger result and could never prove $\omega_1 \to (\omega+n,\ 4)_2^3$ and for a long time (16) remained the strongest result. It was conjectured that perhaps for every ordinal $\alpha < \omega_1$ and integer $n$, $\lambda \to (\alpha, n)_2^3$ and $\omega_1 \to (\alpha, n)_2^3$ holds. Very recently Milner and Prikry using some recent results of Todorcevic proved

$$\omega_1 \to (\omega.2+1, 4)^3_2 .$$

Thus most of the problems here are still open after more than 30 years. For the order types $\lambda$ and $\omega_1$ there are no interesting positive results if we split the four-tuples.

I talked about our results on partition calculus in October 1953 at a meeting of the American Mathematical Society in New York. I stated many of our theorems and open problems and deplored that our results did not find any applications in other branches of mathematics. In the meantime the situation greatly improved. As far as I know Hajnal and Juhász were the first to use partition calculus to solve problems in set-theoretical topology and among them, a problem of de Groot. Juhász tells me that ramification systems were used by Alexandroff and Urysohn in the 1920's. Various results of partition calculus and $\Delta$-systems were used in mathematical logic, and $\Delta$-systems were often used in combinatorics and also occasionally in number theory. In fact the conjecture [5] was discovered because of applications on the greatest prime factors of polynomials and also on problems of combinatorial number theory.

Incidentally Rado and I were the first to prove that for every infinite cardinal number $m$ there exists a graph $G$ of power $m$ and chromatic number $m$ which contains no triangle ([VII], [X]). In our proof we used the construction of Specker by which he proved $\omega^3 \not\to (\omega^3, 3)^2_2$. Perhaps I should mention here a few problems and results on chromatic graphs. The chromatic number $k$ of a graph $G$ is the smallest integer so that the vertices of $G$ can be colored by $k$ colors so that two vertices of the same color are not joined. At first mathematicians became interested in the chromatic number of graphs because of the four-color theorem (at that time it was the four-color problem), but soon it was realized that there are many interesting problems on the chromatic number of graphs which are independent of the four-color problem. Tutte was the first to prove that for every $k$ there is a graph $G$ which has no triangle and which has chromatic number $\geqslant k$ (Ungár, Zykov and Mycielski obtained the same result independently). Tutte sometimes published his results under the pseudonym Blanche Descartes, and in one of my papers quoting this result I referred to Tutte. Smith wrote me a letter saying that Blanche Descartes will be annoyed that I attributed her results to Tutte (he clearly was joking since he knew that I know the facts), but Richard was very precise and when in our paper I wanted to refer to Tutte, Richard only agreed after I got a letter from Smith stating that my interpretation of the facts was correct. After our result with Richard I proved by using the probability method that for every $r$ there is a graph of $n$ vertices the smallest odd circuit of which has size $\geqslant 2r+1$ and whose chromatic number is $> n^{\epsilon_r}$; in fact whose largest independent set is of size $< n^{1-\epsilon_r}$. The order of magnitude of $\epsilon_r$ is known only for $r = 1$.

Here it is known that if $G(n)$ has $n$ vertices and contains no triangle then its chromatic number is less than $\dfrac{n^{1/2}}{(\log n)^{c_1}}$ but can be greater than $\dfrac{n^{1/2}}{(\log n)^{c_2}}$. It would be desirable to get an asymptotic formula.

Some of these theorems could later be proved by constructive methods, the first such proof was due to Lovász and later in a sharper form by Nešetřil and Rödl, but some of the sharper results have not yet been obtained without the probability method.

In a later paper Hajnal and I proved that for every cardinal number $m$ and integer $r$ there is a graph of power $m$, chromatic number $m$ the smallest odd circuit of which has size $\geqslant 2r+1$ and we also proved that every graph of chromatic number $\geqslant \aleph_1$ contains a $k(n;\aleph_1)$ but does not have to contain a $k(\aleph_0, \aleph_0)$

Unfortunately all of us missed the beautiful and fundamental conjecture of Walter Taylor. Let $G$ be any graph of chromatic number $\aleph_1$. Then for every cardinal $m$ there is a graph $G_m$ of chromatic number $m$ for which all finite subgraphs of $G_m$ are subgraphs of $G$ too.

Hajnal, Shelah and I have a triple paper on this subject where we prove some partial results and recently Hajnal and Komjáth [10] have a paper in which they prove many further interesting results on finite and denumerable subgraphs of graphs of chromatic number $\geqslant \aleph_1$. In a triple paper of Hajnal, Szemerédi and myself [11] we prove many interesting theorems and raise many problems which I hope will lead to further interesting results. Thus our old paper with Richard leads to many developments and I am sure will continue to do so.

In [XII], many results are proved and very many unsolved problems are posed but their discussion on the one hand would lead too far into set theory and also a proper discussion of them would need a better knowledge of the many recent results on undecidable problems with which I am not so well acquainted. Hajnal, Shelah and many others could do a better job of this than I. Thus I will restrict myself to a very small sample. Hajnal often observed that to prove positive results in partition calculus we essentially use only two tools. The ramification systems and the canonization Lemma. In its most general form the canonization Lemma is stated in XVIII p. 164 (see also [XII]). This Lemma was one of our most original contributions to set theory and it was very useful in many applications. Shelah has a very significant improvement of our Lemma for $r = 2$ (see p. 159 of [XVIII]). To avoid a complicated formalism we state our Lemma in only a special case: Let $\omega_\alpha$ be a regular cardinal. Let $S_\beta$, $1 \leqslant \beta < \omega_\alpha$, be a rapidly increasing sequence of cardinals. Split the $r$-tuples of $\bigcup_{\beta < \omega_\alpha} S_\beta$ into fewer than $\omega_\alpha$ classes (each $r$-tuple meets each $S_\beta$ in at most one point). Then there is an $A_\beta \subset S_\beta$, $\bigcup_{\beta<\alpha} A_\beta = \bigcup_{\beta<\alpha} S_\beta$, for which the distribution is canonical. In other words if $(x_1 x_2, \ldots, x_n)$ is an $r$-tuple of $\underset{\beta}{U} A_\beta$ and $X_i \in A_{\beta_i}$ then the class of $(x_1, \ldots, x_r)$ only depends on $(\beta_1, \ldots, \beta_r)$. The first triumph of our Lemma was our proof of $\aleph_a \rightarrow [\aleph_a]^3_2$. In human language: If one splits the pairs of a set of power $\aleph_a$ into three classes there always exists a subset of power $\aleph_a$ all pairs of which are in only two of our classes.

To end this paper I just state a random selection of some problems and results which came out of our investigations. Richard and I proved [XIII] that for every pair of integers $m$ and $n$ there is a smallest $l_\alpha(m, n)$ so that

$$\omega_\alpha l_\alpha(m, n) \rightarrow (m, \omega_\alpha n)^2_2 .$$

We conjectured that $l_\alpha(m, n) = l_o(m, n)$ but proved this only for

$m \leqslant 4$, $n \leqslant 2$. Our conjecture was proved by Baumgartner. In [XIII] we also ask: Is it true that $\omega_1 \omega \to (\omega_1 \omega, 3)_2^2$? Hajnal and I later showed

$$\omega_1 \omega \not\to (\omega_1 \omega, 3)_2^2$$

but that

$$\omega_1^2 \to (\omega_1 \omega, 3)_2^2 .$$

We could never decide if $\omega_1^2 \to (\omega_1 \omega, 4)_2^2$ is true. Two years ago Baumgartner and Hajnal proved that the answer is negative. Several unsolved problems remain some of which are discussed in a recent paper of mine. Hajnal and I could not decide whether $\omega_2 \omega \to (\omega_2 \omega, 3)_1^2$ is true. This was recently proved by Shelah and Stanley.

Perhaps I should mention the very useful theorem of Milner and Rado which we used a great deal: Let $\alpha > 0$, $\delta < \omega_{\alpha+1}$. Then

$$\delta \not\to (w_\alpha^{(n)})_{n<\omega}^1 .$$

In human language: One can decompose a well ordered set of order type $\delta < \omega_{\alpha+1}$ into the union of countably many sets each of which has order type $< \omega_\alpha^\omega$. In our triple paper with Milner we used and generalized this important theorem a great deal. Several further problems are stated in papers of Hajnal and myself and Hajnal, Milner and myself.

To end this paper let me state a finite problem: In one of our innumerable mathematical discussions Richard and I observed that if we color the edges of a complete graph $K(m)$ with two colors then in at least one of the colors the resulting graph is connected and contains all the vertices of $K(m)$. This holds both for finite and infinite graphs. The proof is trivial. We wondered what happens if we use more than two colors. This problem was recently considered in several forthcoming papers by Gyárfás and myself but several interesting finite and infinite problems remain which I hope will be further investigated.

## REFERENCES

First I list my joint papers with Rado.


I. "A combinatorial theorem." *Journal of the London Mathematical Society*, vol. 25, 1950, pp. 249-255.

II. "Sets having a divisor property." *American Mathematical Monthly*, vol. 59, 1952, pp. 255-257.

III. "Combinatorial theorems on classifications of subsets of a given set." *Proceedings of the London Mathematical Society*, vol. 2, 1952, pp. 417-439.

IV. "A problem on ordered sets." *Journal of the London Mathematical Society*, vol. 28, 1953, pp. 426-438.

V. "A partition calculus." *Proceedings of the International Congress of Mathematicians*, Amsterdam, 1954, vol. II, p. 55.

VI. "A partition calculus." *Bulletin of the American Mathematical Society*, vol. 62, 1956, pp. 427-489.

VII. "Partition relations connected with the chromatic number of graphs." *Journal of the London Mathematical Society*, vol. 34, 1959, pp. 63-72.

VIII. "A theorem on partial well-ordering of sets of vectors." *Journal of the London Mathematical Society*, vol. 34, 1959, pp. 222-224.

IX. "Intersection theorems for systems of sets." *Journal of the London Mathematical Society*, vol. 35, 1960, pp. 85-90.

X. "A construction of graphs without triangles having pre-assigned order and chromatic number." *Journal of the London Mathematical Society*, vol. 35, 1960, pp. 445-448.

XI. "Intersection theorems for systems of finite sets." *Quarterly Journal of Mathematics*, vol. 12, 1961, pp. 313-320 (with Chao Ko).

XII. "Partition relations for cardinal numbers." *Acta Math. Acad. Scient. Hungarica*, 16, 1965, pp. 93-196 (with A. Hajnal).

XIII. "Partition relations and transitivity domains of binary relations." *Journal of the London Mathematical Society*, 42, 1967, pp. 624-633.

XIV. "Intersection theorems for systems of sets II." *Journal of the London Mathematical Society*, 44, 1969, pp. 467-479.

XV. "Partition relations for *n*-sets." *Journal of the London Mathematical Society (2)*, 3, 1971, pp. 193-204 (with E. Milner).

XVI. "Intersection theorems for systems of sets III." *Journal of the Australian Mathematical Society*, vol. 18, pp. 22-40 (with E. Milner).

XVII. "Transversals and multitransversals." *Journal of the London Mathematical Society (2)*, 19, 1979, pp. 187-191 (with F. Galvin).

XVIII. "Combinatorial set theory: partition relations for cardinals." North-Holland Publishing Co., Hungarian Academy, 1984 (with A. Hajnal and A. Máté).

[1] P. Erdös and A. Hajnal, Unsolved problems in set theory, Axiomatic set theory. Proc. Symp. Pure Math, Vol. XIII, No. I, Amer. Math. Soc., Providence, RI, 1-42. Unsolved and solved problems in set theory, Proc. Tarski, Symp., Proc. Symp. Pure Math, Vol. XXV, Amer. Math. Soc. Providence, RI, 2979, 269-287.

[2] P. Erdös and M. Magidor, A note on regular methods of summability and the Banach-Saks property, Proc. Amer. Math. Soc. 59(1976), 232-234.

[3] P. Erdös and L. Lovász, Problems and results on three chromatic graphs and some related questions, Infinite and Finite Sets, Coll. Keszthely, Hungary, 1973, Coll. Bolyai Math. Soc. 10, 609-624.

[4] P. Erdös and N. Hindman, Enumeration of intersecting epmilies, Dis. Math. 48(1984), 61-65.

[5] For all references see the comprehensive and excellent survey paper of M. Deza and P. Frankl, Erdös-Ko-Rado theorem 22 years later, Siam J. Algebraic and Discrete Methods 4(1983), 419-431, see also R. M. Wilson, The exact bound in the Erdös-Ko-Rado theorem, Combinatorica 4(1984), 247-257.

[6] R. L. Graham, B. Rothschild and J. Spencer, Ramsey Theory, Wiley Interscience Publication in Discrete Math. 1980.

[7] P. Frankl and J. Pach, On disjointly representable sets, Combinatorica 4 (1984), 39-45.

[8] P. Erdös and G. Mills, Some bounds for the Ramsey-Paris-Harrington numbers, J. Comb. Theory A 30(1981), 53-70.

[9] P. Erdös and A. Hajnal, On the structure of set mappings, Acta Math. Acad. Sci. Hungar. 9(1958), 111-130.

[10] A. Hajnal and P. Komjáth, What must and what need not be contained in a graph of uncountable chromatic number, Combinatorica 4 (1984), 47-52.

# The Shifting Technique in Extremal Set Theory

*Peter Frankl*

CNRS
Paris, France

## 1. Introduction and the Erdös-Ko-Rado Theorem.

Let $X$ be a finite set. If not said otherwise we assume $X = \{1, 2, \ldots, n\}$, $n$ a positive integer. For $0 \leqslant k \leqslant n$ we set $2^X = \{F : F \subset X\}$, $\binom{X}{k} = \{F \in 2^X : |F| = k\}$. A *family* $\mathcal{F}$ is just a collection of subsets of $X$, i.e., $\mathcal{F} \subset 2^X$. If $\mathcal{F} \subset \binom{X}{k}$ then it is called *k-uniform*, or a *k-graph*. A family $\mathcal{F}$ is called *intersecting* if $F \cap F' \neq \emptyset$ holds for all $F, F' \in \mathcal{F}$. The simplest intersection theorem is the following:

*Theorem 1.0.* If $\mathcal{F}$ is intersecting then $|\mathcal{F}| \leqslant 2^{n-1}$ holds.

*Proof.* Partition the $2^n$ subsets of $X$ into $2^{n-1}$ pairs where each subset $F$ is paired with its *complement* $X - F$. Since $F \cap (X - F) = \emptyset$, at most one set out of each pair is in $\mathcal{F}$. Thus $|\mathcal{F}| \leqslant 2^{n-1}$. ■

Once an inequality is proved, one is interested in which families attain equality — such families are called *optimal.*

In case of Theorem 1.0 there are many optimal families, in fact, it is not hard to show the following:

*Proposition 1.1.* Given an intersecting family $\mathcal{F} \subset 2^X$, there exists another intersecting family $\mathcal{G} \subset 2^X$, satisfying $\mathcal{F} \subset \mathcal{G}$ and $|\mathcal{G}| = 2^{n-1}$. ■

The first intersection theorem was proved by Erdös, Ko and Rado in the late 1930s, however it was published only in 1961. Before giving its statement one more definition: A family $\mathcal{F}$ is called $t$-intersecting ($t \geqslant 1$, integer) if $|F \cap F'| \geqslant t$ holds for all $F, F' \in \mathcal{F}$.

*Theorem 1.1. (The Erdös-Ko-Rado theorem, [EKR]).* Given $n \geqslant k \geqslant t > 0$ and a $t$-intersecting family $\mathcal{F} \subset \binom{X}{k}$ then for $n \geqslant n_0(k, t)$,

$$|\mathcal{F}| \leqslant \binom{n-t}{k-t} \text{ holds.} \tag{1}$$

To see that the inequality (1) is best possible, i.e., $\max |\mathcal{F}| \geqslant \binom{n-t}{k-t}$, consider the family consisting all $k$-subsets of $X$ which contain $t$ fixed elements. For $n < (k-t+1)(t+1)$ a larger $t$-intersecting family was constructed in [F1] and [EKR]. (See next page). Denote by $n_0(k, t)$ the *least* integer such that (1) holds.

For $n < 2k$ any two $k$-subsets have nonempty intersection, that is $\binom{X}{k}$ is intersecting; and it was shown in [EKR] that $n_0(k, 1) = 2k$. Hilton and Milner [HM] proved that for $t = 1$ and $n > 2k$ the optimal family is unique.

In fact we have $n_0(k, t) = (k-t+1)(t+1)$ for all $t$ and $k$ as it was proved for $t \geqslant 15$ in [F1], and for all $t$ by Wilson [W]. Moreover for $n > n_0(k, t)$ there is only one optimal family.

However, for $t \geqslant 2$ one may ask, what is the maximum size of a $t$-intersecting family $\mathcal{F}$, $\mathcal{F} \subset \binom{X}{k}$ for $2k - t < n < n_0(k, t)$. Denote this maximum by $m(n, k, t)$.

For $0 \leqslant i \leqslant (n-t)/2$ define the family

$$\mathcal{A}_i = \{A \in \binom{X}{k} : |A \cap \{1, 2, \dots, t+2i\}| \geqslant t+i\}.$$

Clearly, $\mathcal{A}_i$ is $t$-intersecting. One can also check that $|\mathcal{A}_1| \gtreqless |A_0| = \binom{n-t}{k-t}$ according as $n \lesseqgtr (k-t+1)(t+1)$.

*Conjecture 1.2. [F1]*

$$m(n, k, t) = \max_i |\mathcal{A}_i|.$$

In [F1] it is shown that for $t \geqslant 15$ and $0.8(k-t+1)(t+1) < n < (k-t+1)(t+1)$ the conjecture is true and $\mathcal{A}_1$ is the only optimal family (up to permutation of the elements). In the case $n = 4n_0$, $k = 2n_0$, $t = 2$ the above conjecture reduces to

$$m(4n_0, 2n_0, 2) \leqslant |\mathcal{A}_{n_0-1}| = \left|\left\{F \in \binom{X}{2n_0} : |F \cap \{1, 2, \dots, 2n_0\}| \geqslant n_0 + 1\right\}\right|.$$

This was already conjectured in [EKR]; however, it appears to be very difficult.

## 2. SHIFTING

Sets have little structure, and this often makes it hard to deal with them. For certain kinds of extremal problems shifting permits one to overcome this difficulty.

For integers $1 \leqslant i < j \leqslant n$ and a family $\mathcal{F}$ define the $(i, j)$-shift $S_{ij}$ as follows

$$S_{ij}(F) = \begin{cases} (F - \{j\}) \cup \{i\} \text{ if } i \notin F,\ j \in F,\ ((F - \{j\}) \cup \{i\}) \notin \mathscr{F} \\ F \text{ otherwise} \end{cases};$$

$$S_{ij}(\mathscr{F}) = \{S_{ij}(F) : F \in \mathscr{F}\} .$$

The next proposition collects some easy but important properties of shifting.

*Proposition 2.1.*

(i) $|S_{ij}(\mathscr{F})| = |\mathscr{F}|$

(ii) If $\mathscr{F}$ is $k$-uniform then so is $S_{ij}(\mathscr{F})$

(iii) If $\mathscr{F}$ is $t$-intersecting then so is $S_{ij}(\mathscr{F})$.

*Proof.* (i) and (ii) are evident. To prove (iii) choose $A_1, A_2 \in S_{ij}(\mathscr{F})$. Let $B_1$, $B_2$ be the corresponding sets in $\mathscr{F}$, i.e., $S_{ij}(B_\nu) = A_\nu$ for $\nu = 1, 2$. Since $|A_1 \cap A_2| \geqslant |B_1 \cap B_2|$ would imply $|A_1 \cap A_2| \geqslant t$, we may assume $|A_1 \cap A_2| < |B_1 \cap B_2|$. This implies $j \in B_1 \cap B_2$ and $\{i, j\} \cap A_1 \cap A_2 = \varnothing$. Say $j \notin A_1$. Then $A_1 = S_{ij}(B_1) = (B_1 - \{j\}) \cup \{i\}$. On the other hand $A_2 = B_2$. Why did we not shift $B_2$ when $i \notin B_2$ and $j \in B_2$? The only possible reason is $B_3 = (B_2 - \{j\}) \cup \{i\} \in \mathscr{F}$. Consequently, $|A_1 \cap A_2| = |B_1 \cap B_3| \geqslant t$ ■

It is not hard to see that if we keep on shifting then finally we end up with a *stable* or *shifted* family $\mathscr{G}$, i.e. $S_{ij}(\mathscr{G}) = \mathscr{G}$ for all $1 \leqslant i < j \leqslant n$. Let us show that $\binom{n}{2}$ shiftings are sufficient if we do them in the right order.

To do this let us first take a different look at shifting.

For $1 \leqslant i \leqslant n$ and a family $\mathscr{F} \subset 2^X$ define $\mathscr{F}(i) = \{F - \{i\}: i \in F \in \mathscr{F}\}$. Then $S_{ij}(\mathscr{F})$ is the unique family $\mathscr{G}$ satisfying $\mathscr{G}(i) = \mathscr{F}(i) \cup \mathscr{F}(j)$, $\mathscr{G}(j) = \mathscr{F}(i) \cap \mathscr{F}(j)$ and $H \in \mathscr{F}$ if and only if $H \in \mathscr{G}$ whenever $|H \cap \{i,j\}| \neq 1$.

Now, it is easy to see that $\mathscr{F}$ is shifted if and only if $\mathscr{F}(j) \subset \mathscr{F}(i)$ holds for all $1 \leqslant i < j \leqslant n$. Since we shall never use Proposition 2.2, its proof will be somewhat sketchy.

*Proposition 2.2.* Let $\mathscr{F} \subset 2^X$ be a family and suppose that we perform in succession all $\binom{n}{2}$ shifts $S_{ij}$, $1 \leqslant i < j \leqslant n$ exactly once, in an order where $S_{ij}$ precedes $S_{i'j'}$ whenever $j' < j$. Then the resulting family is shifted.

*Proof.* Apply induction on $n$. The statement is trivial for $n \leqslant 2$. By the assumptions the $n-1$ shifts $S_{in}$, $1 \leqslant i < n$ are performed first. Let $\mathscr{G}$ be the family after these shifts. Then $\mathscr{G}(n) \subset \mathscr{G}(i)$ can be checked easily. Moreover, this property is maintained during later shifts.

Set $\mathscr{G}(\bar{n}) = \{G \in \mathscr{G}: n \notin \mathscr{G}\}$.

The remaining $\binom{n-1}{2}$ shifts transform $\mathcal{G}(n)$ and $\mathcal{G}(\bar{n})$ independently and the statement follows by induction. ■

*Proposition 2.3.* Suppose $\mathcal{F}$ is $k$-uniform, $t$-intersecting and shifted. Then for all $F_1, F_2 \in \mathcal{F}$

$$|F_1 \cap F_2 \cap [1, 2k-t]| \geqslant t \quad \text{holds.}$$

*Proof.* Take a counter-example maximizing $|F_1 \cap [1, 2k-t]|$. Since $\mathcal{F}$ is $t$-intersecting there is some $j \in (F_1 \cap F_2)$, $j > 2k - t$. Thus $F_1 \cup F_2 \not\subset [1, 2k-t]$, hence we may choose $i \notin F_1 \cup F_2$, $i \leqslant 2k-t$ and replace $F_1$ by $(F_1 - \{j\}) \cup \{i\}$ (recall that $\mathcal{F}$ is shifted), to obtain a contradiction with the maximal choice of $|F_1 \cap [1, 2k-t]|$. ■

Let us now prove the Erdös-Ko-Rado theorem for $t = 1$, $n \geqslant 2k$. Apply induction on $k$ — the statement is trivial for $k = 1$.

a) $n = 2k$. If $F \in \mathcal{F}$ then $|X - F| = n - k = k$, and $(X - F) \notin \mathcal{F}$. Thus $|\mathcal{F}| \leqslant \frac{1}{2}\binom{2k}{k} = \binom{2k-1}{k-1}$, as desired.

b) $n \geqslant 2k$. In view of Proposition 2.1, we may assume that $\mathcal{F}$ is shifted. Define $\mathcal{F}_i = \{F \cap [1, 2k] : F \in \mathcal{F}, |F \cap [1, 2k]| = i\}$.

In view of Proposition 2.3 $\mathcal{F}_i$ is intersecting. By induction $|\mathcal{F}_i| \leqslant \binom{2k-1}{i-1}$ for $i = 0, \ldots, k-1$, the same holds for $i = k$ by a). Given $G \in \mathcal{F}_i$ there are at most $\binom{n-2k}{k-i}$ sets $F \in \mathcal{F}$ with $F \cap [1, 2k] = G$. We infer that

$$|\mathcal{F}| \leqslant \sum_{1 \leqslant i \leqslant k} |\mathcal{F}_i| \binom{n-2k}{k-i} \leqslant \sum_{1 \leqslant i \leqslant k} \binom{2k-1}{i-1}\binom{n-2k}{k-i} = \binom{n-1}{k-1} \quad ■$$

Let us say that the Erdös-Ko-Rado theorem is true for $(n, k, t)$ if $\binom{n-t}{k-t}$ is the maximum size of all $t$-intersecting families $\mathcal{F} \subset \binom{X}{k}$.

*Proposition 2.4.* Suppose the Erdös-Ko-Rado theorem is true for $(n_0, j, t)$, $n_0, t$ fixed and all $j$, $t \leqslant j \leqslant k$. Then it holds for $(n, k, t)$ for all $n > n_0$.

*Proof.* Let us suppose $\mathcal{F}$ is a $t$-intersecting family of maximum size, $\mathcal{F} \subset \binom{X}{k}$, $\mathcal{F}$ is *stable*. For $i \leqslant k$ define $\mathcal{F}_j = \{F \cap [1, n_0] : F \in \mathcal{F}, |F \cap [1, n_0]| = j\}$. In view of Proposition 2.3 $\mathcal{F}_j$ is $t$-intersecting and thus $|\mathcal{F}_j| \leqslant \binom{n_0 - t}{j-t}$ holds. This implies that

$$|\mathcal{F}| \leqslant \sum_{t \leqslant j \leqslant k} |\mathcal{F}_j| \binom{n-n_0}{k-j} \leqslant \sum_{0 \leqslant i \leqslant k-t} \binom{n_0-t}{i} \binom{n-n_0}{k-t-i} = \binom{n-t}{k-t} \blacksquare$$

*Proposition 2.5.* Let $\mathcal{G}$ be a shifted $t$-intersecting family. Then for each $G \in \mathcal{G}$ there exists $i = i(G)$ so that $|G \cap \{1, 2, \ldots, t + 2i\}| \geqslant t + i$.

*Proof.* Let $G = \{x_1, x_2, \ldots, x_{r+t}\}$ with $x_1 < x_2 < \cdots < x_{r+t}$. Suppose that the proposition is not true for $\mathcal{G}$. Note that $r \geqslant 0$ since $\mathcal{G}$ is $t$-intersecting and $x_t \geqslant t + 1$, $x_{t+1} \geqslant t + 3, \ldots, x_{t+r} \geqslant t = 2r + 1$. Adding the trivial $x_1 \geqslant 1, \ldots, x_{t-1} \geqslant t - 1$ and using that $\mathcal{G}$ is shifted, we infer

$$G_1 = \{1, 2, \ldots, t-1, t+1, t+3, \ldots, t+2r+1\} \in \mathcal{G} .$$

Using shiftedness again, it follows that

$$G_2 = \{1, 2, \ldots, t-1, t, t+2, \ldots, t+2r\} \in \mathcal{G} .$$

However, $|\mathcal{G}_1 \cap \mathcal{G}_2| = t - 1$, a contradiction. $\blacksquare$

Let us give now a geometric interpretation of both shifting and this proposition. For this we associate a walk in the plane with each subset $F$ of $\{1, 2, \ldots, n\}$.

We start from the origin and at the $i$-th step we move one unit up if $i \in F$ and one step to the right if $i \notin F$, $1 \leqslant i \leqslant n$. Let us note that this defines a 1–1 correspondence between $2^X$ and all walks of length $n$. For a set $F$ (a walk $w$) let $F(w)$ $(w(F))$ be the corresponding walk (set), respectively. It is clear that if $i \notin F$, $(i + 1) \in F$, then replacing $i + 1$ by $i$ will change the corresponding part of the walk from $\_\!\rfloor$ to $\ulcorner\!\!\!\!-$ . Therefore if $G$ can be obtained from $F$ by such shifts then $w(G)$ is lying above $w(F)$. It is easy to see that for stable families $F \in \mathcal{F}$, $w(G)$ and $w(F)$ end in the same point and $w(G)$ lies above $w(F)$ imply $G \in \mathcal{F}$.

Let us now give a geometric proof of Proposition 2.5. Draw the line $y = x + t$. The integer points of this line have the form $(i, i + t)$. If a walk has some point above this line then it ought to have a point on the line too. But how do we get to the point $(i, i + t)$? Only if our set contains $i + t$ out of the first $2i + t$ elements. Thus to prove Proposition 2.5 we have to show that for a stable, $t$-intersecting family $\mathcal{F}$ none of the walks $w(F)$, $F \in \mathcal{F}$ lies entirely under the line. However, there is a unique highest walk under $y = x + t$; this corresponds to the set $H_1 = \{1, 2, \ldots, t-1, t+1, t+3, \ldots, t+2s+1, \ldots\}$ Therefore if $w(F)$ is under $y = x + t$ for some $F \in \mathcal{F}$ then some subset $F_1$ of $H_1$ is in $\mathcal{F}$, too. Using stability we infer that $F_2 \in \mathcal{F}$ for some $F_2 \subset H_2 = \{1, 2, \ldots, t, t+2, \ldots, t+2s, \ldots\}$. Since $H_1 \cap H_2 = \{1, 2, \ldots, t-1\}$, $|F_1 \cap F_2| \leqslant t - 1$, a contradiction.

## 3. SHADOWS: THE KRUSKAL-KATONA THEOREM

Suppose $\mathcal{F}$ is a family of $k$-element sets, $|\mathcal{F}| = m$. Note that we do not require $\mathcal{F} \subset \binom{X}{k}$. What is the minimum number of $(k - \ell)$-element sets contained in some member of $\mathcal{F}$, as a function of $k$ and $m$? This problem was solved independently by Kruskal [Kr] and Katona [Ka1] more than 20 years ago. To state this result we need some definitions.

For a family $\mathscr{F}$ define its *$\ell$–th shadow*, $\partial_\ell(\mathscr{F})$, by $\partial_\ell(\mathscr{F}) = \{G : \exists F \in \mathscr{F}, G \subset F, |F - G| = \ell\}$. For $\ell = 1$ we write simply $\partial(\mathscr{F})$.

Let us define a linear order on the $k$-subsets of $\{1, 2, ..., n, ...\}$. We say $F < G$ if $F \neq G$ and $\max\{i : i \in F - G\} < \max\{j : j \in G - F\}$ holds. This ordering is called reverse-lexicographic. Suppose $m$ is a given positive integer. Let us take the family of the first $m$ $k$-element sets in this ordering. Denote it by $\mathscr{R}(k, m)$. From the definition it is clear that for some integer $a_k \geqslant k$, $\mathscr{R}(k, m) \subseteq \binom{\{1, ..., a_k\}}{k}$ holds, and moreover, all the remaining sets in $\mathscr{R}(k, m)$ (if there are any) contain $a_k + 1$. Now there is some $a_{k-1} \geqslant k - 1$ so that for all $G \in \binom{\{1, ..., a_{k-1}\}}{k-1}$ one has $(G \cup \{a_k + 1\}) \in \mathscr{R}(k, m)$. Clearly $a_k > a_{k-1}$, because $\binom{[1, a_k + 1]}{k} \not\subseteq \mathscr{R}(k, m)$. Now the remaining sets in $\mathscr{R}(k, m)$ contain both $a_k + 1$ and $a_{k-1} + 1$. Continuing in this way we find elements $a_k > a_{k-1} > \cdots > a_t \geqslant t \geqslant 1$ so that a set $F$ is in $\mathscr{R}(k, m)$ if and only if $F < \{a_t + 1, ..., a_k + 1\}$ holds. This implies $m = \binom{a_k}{k} + \cdots + \binom{a_t}{t}$. This is called the *$k$–cascade representation of $m$.*

*Proposition 3.1.* (i) Every positive integer $m$ has a unique $k$-cascade representation $m = \binom{a_k}{k} + \cdots + \binom{a_t}{t}$ with $a_k > a_{k-1} > \cdots > a_t \geqslant t \geqslant 1$.

(ii) $\partial_\ell(\mathscr{R}(k, m)) = \mathscr{R}\left(k - \ell, \binom{a_k}{k-\ell} + \cdots + \binom{a_t}{t-\ell}\right)$ (where $\binom{a}{b}$ is understood to be zero for $b < 0$).

We leave the easy proof to the reader.

*Theorem 3.2. (Kruskal-Katona theorem)* Suppose $\mathscr{F}$ is a family of $k$-sets, $|\mathscr{F}| = m$ and $m = \binom{a_k}{k} + \cdots + \binom{a_t}{t}$ is the $k$-cascade representation of $m$. Then for all $\ell$, $1 \leqslant \ell \leqslant k$

(3.1) $$|\partial_\ell(\mathscr{F})| \geqslant \binom{a_k}{k - \ell} + \cdots + \binom{a_t}{t - \ell}$$ holds, or equivalently

(3.2) $$|\partial_\ell(\mathscr{F})| \geqslant |\partial_\ell(\mathscr{R}(k, m))| .$$

Because of the $k$-cascade representation, Theorem 3.2 is often clumsy for applications. Lovász proposed the following weaker, but handier version. Recall that $\binom{x}{a} = \dfrac{x(x-1) \cdot ... \cdot (x - a + 1)}{a!}$ can be defined for all real values of $x$.

*Theorem 3.3. ([L]).* Suppose $\mathcal{F}$ is a family of $k$-sets, $|\mathcal{F}| = m$ and $x \geqslant k$ is defined by $m = \binom{x}{k}$. Then for all $1 \leqslant \ell \leqslant k$ one has

$$|\partial_\ell(\mathcal{F})| \geqslant \binom{x}{k-\ell} \tag{3.3}$$

Following [F2] we give a unified argument yielding both results.

*Proof of Theorems 3.2 and 3.3.* First we note that it is sufficient to settle the case $\ell = 1$ (and then iterate the result $\ell$-times) – this is, in fact, trivial from Proposition 3.1 (ii) and the monotonicity of $\binom{x}{a}$ for $x \geqslant a$, respectively.

Our next observation is that for all $1 \leqslant i < j$ one has

$$\partial(S_{ij}(\mathcal{F})) \subset S_{ij}(\partial(\mathcal{F}))$$

– a fact which can be proved by a simple but somewhat tedious case by case analysis. Therefore, in proving (3.1) and (3.3) we may assume that $\mathcal{F}$ is stable (i.e., $S_{ij}(\mathcal{F}) = \mathcal{F}$ for all $1 \leqslant i < j$). We apply induction on $m$ and for given $m$ on $k$. Note that both statements are trivial for $k = 1$, $m$ arbitrary.

Let us define two new families

$$\mathcal{F}_0 = \{F \in \mathcal{F} : 1 \notin F\}$$

$$\mathcal{F}_1 = \{F - \{1\} : 1 \in F \in \mathcal{F}\} .$$

Clearly,

$$|\mathcal{F}_0| + |\mathcal{F}_1| = |\mathcal{F}| . \tag{3.4}$$

Since $\mathcal{F}$ is stable, we have

$$\partial\mathcal{F}_0 \subset \mathcal{F}_1 . \tag{3.5}$$

We claim

$$|\mathcal{F}_1| \geqslant \binom{x-1}{k-1} . \tag{3.6}$$

In fact, $|\mathcal{F}| = \binom{x}{k} = \binom{x-1}{k} + \binom{x-1}{k-1}$ and (3.4) would imply that if (3.6) were not true then $|\mathcal{F}_0| > \binom{x-1}{k}$, so that by induction $|\partial\mathcal{F}_0| > \binom{x-1}{k-1}$, contradicting (3.5). Therefore (3.6) is true. Now note that $\partial\mathcal{F} \subset \mathcal{F}_1 \cup \{\{1\} \cup G : G \in \partial\mathcal{F}_1\}$. By induction $|\partial\mathcal{F}_1| \geqslant \binom{x-1}{k-2}$, and thus

$$|\partial\mathcal{F}| \geqslant \binom{x-1}{k-1} + \binom{x-1}{k-2} = \binom{x}{k-1}, \text{ proving (3.3) .}$$

To prove (3.1) we first show that one can assume that

$$(3.7) \qquad |\mathcal{F}_1| \geqslant \binom{a_k-1}{k-1} + \cdots + \binom{a_t-1}{t-1}.$$

If this were not the case then (3.4) would imply that

$$(3.8) \qquad |\mathcal{F}_0| \geqslant \binom{a_k-1}{k} + \cdots + \binom{a_t-1}{t} + 1.$$

If $a_t - 1 \geqslant t$, then we can forget about the +1 in (3.8) and deduce from the induction hypothesis that

$$|\partial\mathcal{F}_0| \geqslant \binom{a_k-1}{k-1} + \cdots + \binom{a_t-1}{t-1}, \text{ contradicting (3.5) .}$$

If $a_t = t$ then let $s$ be the largest integer so that $a_s = s$ holds, $k \geqslant s \geqslant t$. Then (3.8) can be rewritten as

$$|\mathcal{F}_0| \geqslant \binom{a_k-1}{k} + \cdots + \binom{a_{s+1}-1}{s+1} + \binom{s}{s}.$$

From the induction hypothesis we infer that

$$\begin{aligned} |\partial\mathcal{F}_0| &\geqslant \binom{a_k-1}{k-1} + \cdots + \binom{a_{s+1}-1}{s} + \binom{s}{s-1} \\ &\geqslant \binom{a_k-1}{k-1} + \cdots + \binom{a_{s+1}-1}{s} + (s-t+1) \\ &= \binom{a_k-1}{k-1} + \cdots + \binom{a_t-1}{t-1}, \end{aligned}$$

again in contradiction to (3.5). Therefore we can assume that (3.7) is true. We conclude the proof of (3.1) as that of (3.3), i.e., using $|\partial\mathcal{F}| \geqslant |\partial\mathcal{F}_1| + |\mathcal{F}_1|$. By the induction hypothesis and (3.7), $|\partial\mathcal{F}_1| \geqslant \binom{a_k-1}{k-2} + \cdots + \binom{a_t-1}{t-2}$. Adding this inequality to (3.7), (3.1) follows. ■

*Proposition 3.4.* Suppose that $\mathcal{F}$ is a family of $k$ sets, $|\mathcal{F}| = m \geqslant 1$ and $x \geqslant k$ is defined by $m = \binom{x}{k}$. Suppose further that $|\partial_\ell(\mathcal{F})| = \binom{x}{k-\ell}$ holds for some $\ell$,

$1 \leqslant \ell < k$. Then $x$ is an integer and $\mathscr{F} = \binom{X_0}{k}$ holds for some $x$-element set $X_0$.

*Proof.* Suppose first that $\ell = 1$ and $\mathscr{F}$ is stable. Recall the proof of Theorem 3.3.

We conclude that equality must hold in (3.6). That is,

$$|\mathscr{F}_1| = \binom{x-1}{k-1}. \tag{3.9}$$

Since $\binom{x}{k} = \frac{x}{k}\binom{x-1}{k-1}$ and both $\binom{x}{k}$ and $\binom{x-1}{k-1}$ are integers, $x$ must be rational. Now the fact, that $\binom{x}{k}$ is an integer, implies that $x$ is an integer too.

Apply induction on $x$. For $x = k$ the statement is trivially true.

From (3.9) we infer that

$$|\mathscr{F}_0| = \binom{x-1}{k}. \tag{3.10}$$

By (3.5) and Theorem 3.3 we infer from (3.9) and (3.10) that

$$\partial\mathscr{F}_0 = \mathscr{F}_1 \text{ and } |\partial\mathscr{F}_0| = \binom{x-1}{k-1}. \tag{3.11}$$

By the induction hypothesis $\mathscr{F}_0 = \binom{Y}{k}$ holds for some set $Y$ with $|Y| = x - 1$.

Now $\mathscr{F} = \binom{Y \cup \{1\}}{k}$ follows from (3.11).

To settle the case $\ell = 1$ we have to deal with non-shifted families as well. By the above argument we may assume that $x$ is an integer. We have to show that $|\cup\mathscr{F}| = x$.

We know, that this must hold for the shifted family. Therefore we may indirectly assume that there exist $1 \leqslant i < j \leqslant n$ such that $|\cup\mathscr{F}| = x + 1$ but $|\cup S_{ij}(\mathscr{F})| = x$.

That is, the $(i, j)$-shift removes all sets containing $j$ from $\mathscr{F}$. Also, $|\mathscr{F}| = \binom{x}{k}$ implies $S_{ij}(\mathscr{F}) = \binom{\cup S_{ij}(\mathscr{F})}{k}$. Set $Y = \cup S_{ij}(\mathscr{F})$ and note $j \notin Y$. Consider the following two families:

$$\mathscr{G} = \{F - \{i\} : i \in F \in \mathscr{F}\}, \quad \mathscr{H} = \{F - \{j\} : j \in F \in \mathscr{F}\}.$$

Then $\mathcal{G} \cap \mathcal{H} = \varnothing$ and $\mathcal{G} \cup \mathcal{H} = \binom{Y-\{i\}}{k-1}$ hold. Since $|\cup \mathcal{F}| = x + 1$, $\mathcal{G} \neq \varnothing \neq \mathcal{H}$ follows. Consequently, there exists some $B \in \binom{Y-\{i\}}{k-2}$ with $B \in (\partial\mathcal{G} \cap \partial\mathcal{H})$.

Therefore $B \cup \{j\}$ does not change when one applies the $(i, j)$-shift to $\partial\mathcal{F}$. Thus

$$\partial(S_{ij}(\mathcal{F})) \subsetneq S_{ij}(\partial(\mathcal{F})) \text{ holds, i.e., } |\partial(\mathcal{F})| > \binom{x}{k-1}.$$

Let now $\ell \geqslant 2$. If $|\partial_1(\mathcal{F})| = \binom{x}{k-1}$, then we are done by the preceding case. If $|\partial_1(\mathcal{F})| = \binom{y}{k-1}$ where $y > x$, then Theorem 3.3 implies $|\partial_\ell(\mathcal{F}) = \partial_{\ell-1}(\partial(\mathcal{F}))| \geqslant \binom{y}{k-\ell}$, a contradiction. ■ Let us mention that recently Füredi-Griggs [FG] and Mörs [M] characterized those triples $(m, k, \ell)$ for which $\mathcal{R}(k, m)$ is the unique optimal family in the Kruskal-Katona Theorem.

*Corollary 3.5. (Sperner [S])* Suppose that $\varnothing \neq \mathcal{F} \subset \binom{X}{k}$. Then $\partial_\ell(\mathcal{F})/|\mathcal{F}| \geqslant \binom{n}{k-\ell} / \binom{n}{k}$ holds with equality if and only if $\mathcal{F} = \binom{X}{k}$.

*Proof.* Note that $\binom{x}{k-\ell} / \binom{x}{k} = k!/(k-\ell)!(x-k+\ell) \cdot \ldots \cdot (x-k+1)$ is monotone decreasing and apply Theorem 3.3 together with Proposition 3.4. ■

Note that this corollary can be easily proved by a direct double-counting argument, too.

## 4. SHADOWS OF $t$-INTERSECTING FAMILIES

If one assumes that $\mathcal{F} \subset \binom{X}{k}$ is $t$-intersecting then the bound of the Kruskal-Katona theorem for $|\partial_\ell \mathcal{F}|$ can be improved. In particular, we shall show $|\partial_\ell \mathcal{F}| \geqslant |\mathcal{F}|$ for $\ell \leqslant t$.

Let us first consider $\mathcal{A} = \binom{[1,2k-t]}{k}$. Clearly $\mathcal{A}$ is $t$-intersecting and $|\partial_\ell \mathcal{A}| = \binom{2k-t}{k-\ell}$. The next theorem shows that $\mathcal{A}$ is the "worst example".

*Theorem 4.1 (Katona [Ka2])* Suppose that $\mathcal{F}$ is a $k$-uniform, $t$-intersecting family. Then for $1 \leqslant \ell \leqslant t$

$$(4.1) \qquad |\partial_\ell \mathscr{F}| \geqslant |\mathscr{F}| \frac{\binom{2k-t}{k-\ell}}{\binom{2k-t}{k}}$$

holds.

*Proof of Theorem 4.1. (cf. [F3]).* In view of Propositions 2.1 and 2.4 we may assume that $\mathscr{F}$ is shifted. Then in view of Proposition 2.5 for each $F \in \mathscr{F}$ there exists $i$ so that $|[1, t+2i] \cap F| \geqslant t+i$ holds. Let $i(F)$ denote the maximum value of $i$ for which this holds, i.e., for all $j > i(F)$ one has $|F \cap [1, t+2j]| < t+j$ and, consequently, $|F \cap [1, t+2i(F)]| = t+i(F)$. This makes it possible to partition $\mathscr{F}$ according to $i(F)$ and $F \cap [t+2i(F)+1, n]$.

First define for $A \in \binom{[2t+i+1,n]}{k-t-i}$, $\mathscr{F}_A = \{F \in \mathscr{F} : i(F) = i, F \cap [2t+i+1, n] = A\}$. Then we have the partition

$$\mathscr{F} = \bigcup_{0 \leqslant i \leqslant k-t} \quad \bigcup_{A \in \binom{[2t+i+1,n]}{k-t-i}} \mathscr{F}_A .$$

Define $\overline{\mathscr{F}}_A = \{F - A : F \in \mathscr{F}_A\}$ and note that $\overline{\mathscr{F}}_A \subset \binom{[1, t+2i]}{t+i}$. Thus by Corollary 3.5 and the fact that $\binom{t+2i}{t+i-\ell} / \binom{t+2i}{t+i}$ is monotone increasing as a function of $i$ for fixed $0 \leqslant \ell \leqslant t$ we have

$$(4.2) \qquad |\partial_\ell \overline{\mathscr{F}}_A| \geqslant |\overline{\mathscr{F}}_A| \binom{t+2i}{t+i-\ell} / \binom{t+2i}{t+i} \geqslant |\overline{\mathscr{F}}_A| \binom{2k-t}{k-\ell} / \binom{2k-t}{k} .$$

Define $\tilde{\partial}_\ell \mathscr{F}_A = \{G \cup A : G \in \partial_\ell \overline{\mathscr{F}}_A\}$. It is immediate that $\tilde{\partial}_\ell \mathscr{F}_A \subset \partial_\ell \mathscr{F}_A$ holds. We claim that for $A$, $A'$ distinct $\tilde{\partial}_\ell \mathscr{F}_A \cap \tilde{\partial}_\ell \mathscr{F}_{A'} = \varnothing$.

Suppose $|A| = k - t - i$, $|A'| = k - t - i'$, $i \leqslant i'$. Let us first consider the case $i = i'$. Since for $H \in \tilde{\partial}_\ell \mathscr{F}_A (\tilde{\partial}_\ell \mathscr{F}_{A'})$ one has $H \cap [t + 2i + 1, n] = A (A')$, respectively, we see that the same $H$ cannot be in both families.

Suppose next $i < i'$, $H \in \tilde{\partial}_\ell \mathscr{F}_A$, $H' \in \tilde{\partial}_\ell \mathscr{F}_{A'}$, let $F, F'$ be respective members of $\mathscr{F}_A$, $\mathscr{F}_{A'}$ satisfying $H \subset F$, $H' \subset F'$. Note that $i(F) = i$, $i(F') = i'$. Thus the definition of $i(F)$ implies $|F \cap [1, 2i' + t]| < i' + t$, $|F' \cap [1, 2i' + t]| = i' + t$. Consequently $|H \cap [1, 2i' + t]| < i' + t - \ell = |H' \cap [1, 2i' + t]|$, showing $H \neq H'$. Therefore summing (4.2) over all $0 \leqslant i \leqslant k - t$ and all $A \in \binom{[2i+t+1, n]}{k-t-i}$ the inequality (4.1) follows. ■

*Remark 4.3.* One can show, using the above approach, that in (4.1) equality holds only for $\mathcal{F} = \binom{[1, 2k-t]}{k}$. Moreover, the following result of Füredi and the author can be deduced.

*Theorem 4.4.* Suppose $\mathcal{F}$ is $k$-uniform, $t$-intersecting and $|\mathcal{F}| > m_0(k, t)$. Then for $1 \leqslant \ell < t$,

$$|\partial_\ell \mathcal{F}|/|\mathcal{F}| > \binom{2k-2-t}{k-1-\ell} / \binom{2k-2-t}{k-1}. \tag{4.3}$$

The inequality (4.3) is asymptotically best possible as is seen by considering $\mathcal{A}_{k-t-1} = \left\{F \in \binom{[1,n]}{k} : |F \cap [1, 2k-2-t]| \geqslant k-1\right\}$, $n$ tending to infinity.

## 5. THE MAXIMUM SIZE OF $t$-INTERSECTING FAMILIES: THE KATONA-THEOREM

In Theorem 1.0 we showed that $2^{n-1}$ is the maximum size of an intersecting family $\mathcal{F} \subset 2^{[1,n]}$. What if $\mathcal{F}$ is $t$-intersecting? Let us first give examples of large $t$-intersecting families:

$$\mathcal{B}(n, t) = \left\{B \in X : |B| \geqslant \frac{n+t}{2}\right\}.$$

It is clear that $\mathcal{B}(n, t)$ is $t$-intersecting. However, for $n + t$ odd one can add $\frac{n+t-1}{2}$-element sets to $\mathcal{B}(n, t)$ and still have a $t$-intersecting family.

$$\mathcal{B}^*(n, t) = \mathcal{B}(n, t) \cup \left\{B \in \binom{X}{\frac{n+t-1}{2}} : n \notin B\right\}.$$

It is easy to check that $\mathcal{B}^*(n, t) = \{B \subset X : B \cap [1, n-1] \in \mathcal{B}(n-1, t)\}$. Thus $|\mathcal{B}^*(n, t)| = 2|\mathcal{B}(n-1, t)|$ holds.

*Theorem 5.1. (Katona ([Ka2]))* Suppose $\mathcal{F} \subset 2^X$ is $t$-intersecting. Then one of the following two cases occurs.

a) $\quad n + t = 2s$ and $|\mathcal{F}| \leqslant |\mathcal{B}(n, t)| = \sum_{i=s}^{n} \binom{n}{i}$

b) $\quad n + t = 2s + 1$ and $|\mathcal{F}| \leqslant |\mathcal{B}^*(n, t)| = \binom{n-1}{s} + \sum_{i=s+1}^{n} \binom{n}{i}$.

Moreover, for $t \geqslant 2$ the only optimal families are $\mathcal{B}(n, t)$ and $\mathcal{B}^*(n, t)$, respectively.

*Proof.* First note that $|F \cap F'| \geqslant t$ implies that for $G \subset F$, $|F - G| = t - 1$ one still has $G \cap F' \neq \varnothing$. In particular, no member of $\partial_{t-1}\mathscr{F}$ can be the complement of a member of $\mathscr{F}$. To apply this observation define $\mathscr{F}^{(i)} = \{F \in \mathscr{F}: |F| = i\}$, $f_i = |\mathscr{F}^{(i)}|$. Then we have

$$|\partial_{t-1}\mathscr{F}^{(i)}| + |\mathscr{F}^{(n+t-1-i)}| \leqslant \binom{n}{n+t-1-i}, \quad t \leqslant i \leqslant \frac{n+t-1}{2}$$

Using (4.1) we infer that

$$\text{(5.1)} \qquad \frac{i}{i-t+1} f_i + f_{n+t-1-i} \leqslant \binom{n}{n+t-1-i}, \quad t \leqslant i \leqslant \frac{n+t-1}{2}.$$

Summing (5.1) for $t \leqslant i < \dfrac{n+t-1}{2}$ and noting $f_n \leqslant 1$, $f_i = 0$ for $i < t$ we obtain in the case $n + t = 2s$

$$|\mathscr{F}| = \sum_{i=0}^{n} f_i \leqslant \left[\sum_{i=t}^{s-1} \frac{i}{i-t+1} f_i + f_{n+t-1-i}\right] + f_n \leqslant \sum_{j=s}^{n} \binom{n}{j}, \text{ proving}$$

the theorem for this case. If $t \geqslant 2$, then $i > i - t + 1$. Thus to have equality, one must have $f_i = 0$ for $i \leqslant s - 1 = \dfrac{n+t-2}{2}$, i.e., $\mathscr{F} \subset \mathscr{B}(n, t)$.
In the case $n + 1 = 2s + 1$, we obtain in the same way

$$|\mathscr{F}| \leqslant |\mathscr{B}(n, t)| + |\mathscr{F}^{(s)}|.$$

Applying (5.1) for $i = s = \dfrac{n+t-1}{2}$ gives $|\mathscr{F}^{(s)}| = f_s \leqslant \dfrac{n-s}{n}\dbinom{n}{s} = \dbinom{n-1}{s}$, yielding $|\mathscr{F}| \leqslant |\mathscr{B}^*(n, t)|$, as desired. The uniqueness of the optimal family can be argued similarly to case a). ■

Let us note that for $F, F' \subset X$ $|F \cap F'| \geqslant t$ is equivalent to $|F \cup F'| \leqslant n - t$. Thus the Katona Theorem can be restated as follows.

*Theorem 5.2.* Suppose that $\mathscr{F} \subset 2^X$ satisfies for all $F, F' \in \mathscr{F}$,

$$\text{(5.1)} \qquad |F \cup F'| \leqslant b < n.$$

Then one of the following holds

$$\text{(i)} \qquad b = 2k \text{ and } |\mathscr{F}| \leqslant \sum_{i=0}^{k} \binom{n}{i},$$

$$\text{(ii)} \qquad b = 2k - 1 \text{ and } |\mathscr{F}| \leqslant \binom{n-1}{k-1} + \sum_{0 \leqslant i \leqslant k} \binom{n}{i}$$

Moreover, for $b \leqslant n - 2$, the optimal families are unique.

Note that in case (ii) for every intersecting family $\mathcal{G} \subset \binom{X}{k}$ the family $\mathcal{G} \cup \{F \subset X : |F| < k\}$ satisfies (5.1). Thus Theorem 5.2(ii) implies that $|\mathcal{G}| \leqslant \binom{n-1}{k-1}$, which is the Erdös-Ko-Rado Theorem. Also, the uniqueness of the optimal families implies that for $n > 2k$ there is a unique optimal family in the Erdös-Ko-Rado Theorem, as well.

## 6. THE HILTON-MILNER THEOREM.

As we saw in the preceding section, the Katona Theorem implies the Erdös-Ko-Rado Theorem together with the uniqueness of the optimal families for $n > 2k$. Hilton and Milner described the next two optimal families.

Define $\mathcal{H} = \left\{H \in \binom{[1,n]}{k} : 1 \in H,\ [2, k+1] \cap H \neq \emptyset\right\} \cup \{[2, k+1]\}$.

Clearly, $\mathcal{H}$ is intersecting and $\cap \mathcal{H} = \emptyset$. Define also $\mathcal{G} = \left\{G \in \binom{[1,n]}{k} : |G \cap [1,3]| \geqslant 2\right\}$. Note that $\mathcal{G}$ is intersecting, $\cap \mathcal{G} = \emptyset$ and for $k = 2$, $\mathcal{G} = \mathcal{H}$ holds.

*Theorem 6.1. (Hilton-Milner Theorem ([HM]))* Suppose that $\mathcal{F} \subset \binom{X}{k}$ is intersecting, $n > 2k$ and $\cap \mathcal{F} = \emptyset$. Then

$$(6.1) \qquad |\mathcal{F}| \leqslant |\mathcal{H}| = \binom{n-1}{k-1} - \binom{n-k-1}{k-1} + 1 .$$

Moreover, equality holds in (6.1) if and only if $\mathcal{F}$ is isomorphic to $\mathcal{H}$, or $k = 3$ and $\mathcal{F}$ is isomorphic to $\mathcal{G}$.

The original proof of this theorem is rather involved. For other proofs cf. Mörs [M] and Alon [A]. The present proof is due to [FF].

*Proof.* We start by applying the $(i, j)$-shift to $\mathcal{F}$. Then either $S_{ij}(\mathcal{F})$ satisfies the assumptions of the theorem or $i$ is contained in every member of $S_{ij}(\mathcal{F})$. In the first case we keep on shifting until, eventually, we obtain a shifted family satisfying the assumptions.

Suppose now that at some point the second possibility occurs. Without loss of generality suppose $i = 1$, $j = 2$. Since $1 \in F$ for all $F \in S_{ij}(\mathcal{F})$, $\{1, 2\}$ intersects all members of $\mathcal{F}$. Taking $\mathcal{F}$ of maximal size we may assume that

$$(6.2) \qquad \left\{G : [1,2] \subset G \in \binom{X}{k}\right\} \subset \mathcal{F} .$$

Since $\cap \mathcal{F} = \emptyset$, we may assume that $\{1, 3, 4, \ldots, k+1\} \in \mathcal{F}$. Now, instead of $S_{12}$ we keep applying the $(i, j)$-shift for $3 \leqslant i < j \leqslant n$. Then (6.2) implies that

$\cap S_{ij}(\mathscr{F}) = \emptyset$. Eventually we obtain a family, which we denote by abuse of notation by $\mathscr{F}$, satisfying $S_{ij}(\mathscr{F}) = \mathscr{F}$ for all $3 \leqslant i < j \leqslant n$. Note that $G_\ell = \{\ell\} \cup [3, k+1]$ is the unique smallest set $G \in \binom{X}{k}$ satisfying $G \cap [1,2] = \{\ell\}$, $\ell = 1, 2$. Thus $\cap \mathscr{F} = \emptyset$ and the stability of $\mathscr{F}$ implies $G_1$, $G_2 \in \mathscr{F}$. Together with (6.2) this yields

$$\binom{[1, k+1]}{k} \subset \mathscr{F}. \tag{6.3}$$

Now we can apply an arbitrary $(i, j)$-shift, even with $i = 1, 2$, and $\binom{[1, k+1]}{k}$ will not change and therefore $\cap \mathscr{F} = \emptyset$ will be maintained.

Consequently, in proving (6.1) we may assume that $\mathscr{F}$ is a stable family. Now stability implies $[2, k+1] \in \mathscr{F}$ and thus (6.3) holds by stability.

We apply induction on $n$. Define $\mathscr{F}_i = \{F \cap [1, 2k] : F \in \mathscr{F}, |F \cap [1, 2k]| = i\}$, $0 \leqslant i \leqslant k$. In view of Proposition 2.3 the family $\bigcup_i \mathscr{F}_i$ is intersecting. Consequently, $\mathscr{F}_0 = \emptyset$. Also, (6.3) implies $\mathscr{F}_1 = \emptyset$.

*Claim 6.2*

$$|\mathscr{F}_i| \leqslant \binom{2k-1}{i-1} - \binom{k-1}{i-1}, \quad 2 \leqslant i < k \text{ and} \tag{6.4}$$

$$|\mathscr{F}_k| \leqslant \binom{2k-1}{k-1} - \binom{k-1}{k-1} + 1 \text{ hold}. \tag{6.5}$$

*Proof.* If $\cap \mathscr{F}_i \neq \emptyset$, then (6.3) implies (6.4). If $\cap \mathscr{F}_i = \emptyset$ then by the induction assumption

$$|\mathscr{F}_i| \leqslant \binom{2k-1}{i-1} - \binom{2k-i-1}{i-1} + 1 \leqslant \binom{2k-1}{i-1} - \binom{k-1}{i-1}, \text{ proving (6.4)}.$$

For $i = k$, $|\mathscr{F}_k| \leqslant \frac{1}{2}\binom{2k}{k} = \binom{2k-1}{k-1} - \binom{k-1}{k-1} + 1$ is trivial because $\mathscr{F}_k$ is intersecting. ∎

Given $A \subset [1, 2k]$, there are at most $\binom{n-2k}{k-|A|}$ sets $F \in \mathscr{F}$ with $F \cap [1, 2k] = A$. Thus Claim 6.2 implies:

$$|\mathscr{F}| \leqslant \sum_{i=1}^{k} \binom{n-2k}{k-i} |\mathscr{F}_i| \leqslant 1 + \sum_{i=2}^{k} \binom{n-2k}{k-i} \left( \binom{2k-1}{i-1} - \binom{k-1}{i-1} \right)$$

$$= 1 + \binom{n-1}{k-1} - \binom{n-k-1}{k-1},$$

proving (6.1).

If we have equality, then $|\mathcal{F}_2| = \max\{3, k\}$ follows. Since $\mathcal{F}_2$ is intersecting, either $\mathcal{F}_2 = \{\{1, j\} : 2 \leqslant j \leqslant k + 1\}$ and consequently, $\mathcal{F} \subset \left\{H \in \binom{X}{k} : H \cap A \neq \emptyset \text{ for all } A \in \mathcal{F}_2\right\} = \mathcal{H}$. Or $k = 3$, $\mathcal{F}_2 = \binom{[1,3]}{2}$, and $\mathcal{F} \subset \{G \in \binom{X}{k} : |G \cap [1, 3]| \geqslant 2\} = \mathcal{G}$ This proves the uniqueness of the optimal-families among stable families.

The general case follows from the fact — whose proof is easy and omitted — that if $\mathcal{F}$ is intersecting and $S_{ij}(\mathcal{F}) = \mathcal{G}$ or $\mathcal{H}$ then $\mathcal{F} = \mathcal{G}$ or $\mathcal{H}$. ■

Let us mention the following sharpening of the Hilton-Milner Theorem — the proof of which uses shifting as well.

For $3 \leqslant i \leqslant k + 1$, define

$$\mathcal{H}_i = \left\{H \in \binom{X}{k} : 1 \in H, [2,i] \cap H \neq \emptyset\right\} \cup \left\{\binom{X}{k} : 1 \notin H, [2,i] \subset H\right\}.$$

Note that $\mathcal{H} = \mathcal{H}_{k+1}$ and $\mathcal{G} = \mathcal{H}_3$. Also for $n > 2k$ $|\mathcal{H}_3| = |\mathcal{H}_4| < |\mathcal{H}_5| < \cdots < |\mathcal{H}_{k+1}|$.

For $\mathcal{F} \subset \binom{X}{k}$ let $d(\mathcal{F})$ be the maximum degree of $\mathcal{F}$, i.e., $d(\mathcal{F}) = \max_{1 \leqslant i \leqslant n} |\{F \in \mathcal{F} : i \in \mathcal{F}\}|$.

*Theorem 6.3 ([F4])* Suppose that $\mathcal{F} \subset \binom{X}{k}$ is intersecting and $d(\mathcal{F}) \leqslant d(\mathcal{H}_i)$ holds for some $3 \leqslant i \leqslant k + 1$. Then $|\mathcal{F}| \leqslant |\mathcal{H}_i|$, moreover, equality holds if and only if either $\mathcal{F}$ is isomorphic to $\mathcal{H}_i$ or $i = 4$ and $\mathcal{F}$ is isomorphic to $\mathcal{H}_3$.

To obtain the Hilton-Milner Theorem just observe that $d(\mathcal{F}) > d(\mathcal{H}_{k+1})$ immediately implies $\cap \mathcal{F} \neq \emptyset$.

## 7. ON $r$-WISE $t$-INTERSECTING FAMILIES

A family $\mathcal{F} \subset 2^X$ is called $r$-wise $t$-intersecting if any $r$ members of it intersect in at least $t$ elements. Denote by $f(n, r, t)$ the maximum size of all $r$-wise $t$-intersecting families in $2^X$.

*Proposition 7.1.*

(i) $$f(n + 1, r, t) \geqslant 2f(n, r, t) \text{ for } n \geqslant t,$$

(ii) $$p(r, t) = \lim_{n \to \infty} f(n, r, t)/2^n \text{ exists for all } r, t.$$

*Proof.*

(i) If $\mathcal{F} \subset 2^X$ is $r$-wise $t$-intersecting then so is

$$\mathcal{F}^* = \{F \subseteq (X \cup \{n+1\}) : F \cap [1,n] \in \mathcal{F}\} .$$

(ii) In view of (i) the function $f(n, r, t)/2^n$ is monotone non-decreasing in $n$ and it is clearly bounded above by 1. ■

Let us note that $p(r, t) \leqslant 1/2$ follows from Theorem 1.0. Considering $\mathcal{B}(n, t)$ from Theorem 5.1 one sees that in fact

$$p(2, t) = \frac{1}{2} \text{ for all } t \geqslant 1 .$$

Soon we will see that the situation is radically different for $r > 2$ and in fact $p(r, t)/p(r-1, t)$ tends to zero exponentially fast for $r$ fixed and $t \to \infty$.

It is easy to see that Proposition 2.1.(iii) holds for $r$-wise $t$-intersecting families; therefore we will assume that $\mathcal{F}$ is a stable, $r$-wise $t$-intersecting family.

*Proposition 7.2.* For each $F \in \mathcal{F}$ there exists some $i \geqslant 0$ so that

(7.1) $$|F \cap [1, t+ri]| \geqslant t + (r-1)i \text{ holds} .$$

*Proof.* To prove (7.1) we apply the geometric approach of the proof of Proposition 2.5.

Then the statement is equivalent to saying that for $F \in \mathcal{F}$ $w(F)$ meets the line $y = t + (r-1)x$. Again, there is a unique maximal walk not meeting this line, corresponding to the (infinite) set

$$A_0 = \{1, 2, \ldots, t-1, t+1, \ldots, t+r-1, t+r+1, \ldots, t+2r-1, t+2r+1, \ldots\},$$

i.e., $A_0$ misses $t,\ t+r,\ t+2r,\ \ldots$ .

If Proposition 7.2. was not true then for some $F \in \mathcal{F}$ $w(F)$ would lie under $w(A_0)$. Using the stability of $\mathcal{F}$ one finds $F_0 \in \mathcal{F}$ satisfying $F_0 \subset A_0$.

Let us define $A_i = \{1, 2, \ldots, n, \ldots\} - \{t+i,\ t+i+r,\ t+i+2r, \ldots\}$, $1 \leqslant i < r$. Since $F_0 \subset A_0$ and $A_i$ can be obtained from $A_0$ by shifting, there exist $F_i \subset A_i$, $F_i \in \mathcal{F}$ for all $0 \leqslant i < r$. However, $|F_0 \cap \cdots \cap F_{r-1}| \leqslant |A_0 \cap \cdots \cap A_{r-1}| = t - 1$, a contradiction. ■

Since a set uniquely corresponds to a (0,1)-vector, its characteristic vector, one can give a probabilistic interpretation for the ratio of sets satisfying (7.1): consider the infinite random walk in which at each step we move one unit with probability $\frac{1}{2}$ up or right. Then the above ratio is the probability $q(r, t)$ that we ever hit the line $y = t + (r-1)x$.

It is easy to see that $q(2, t) = 1$ for all $t$ while $q(,r\,t) < 1$ for all $r \geqslant 3$.

*Proposition 7.3.* Let $\alpha_r$ denote the unique root in the interval $\left(\frac{1}{2}, 1\right)$ of the polynomial $z^r - 2z + 1$. Then

(i) $$q(r, t) = \alpha_r^t .$$

(ii) $$\frac{1}{2} < \alpha_r < \frac{1}{2} + \frac{1}{2^r}$$

*Proof.* First note that $q(r, t)$ satisfies the linear recursion $q(r, 0) = 1$, $q(r, t + 1) = \frac{1}{2}q(r, t) + \frac{1}{2}q(r, t + r)$. Next we prove that $q(r, t)$ is multiplicative, i.e.,

(7.2) $$q(r, t + s) = q(r, t)q(r, s) .$$

Let $q_i$ be the probability that a walk hits the line $y = t + s + (r - 1)x$ and the first place it hits $y = t + (r - 1)x$ is at $x = i$. Obviously

(7.3) $$q(r, t + s) = \sum_{i \geqslant 0} q_i .$$

Let $p_i$ be the probability that a walk hits the line $y = t + (r - 1)x$ first in the place $x = i$. Clearly

$$q(r, t) = \sum_{i \geqslant 0} p_i \text{ and } q_i = p_i q(r, s) .$$

Thus (7.3) implies

$$q(r, t + s) = \sum_{i \geqslant 0} p_i q(r, s) = q(r, t)q(r, s), \text{ proving (7.2)} .$$

In view of (7.2) $q(r, t) = q(r, 1)^t$ holds. Let us set $\beta_r = q(r, 1)$. Substituting this into the recurrence relation $2\beta_r = 1 + \beta_r^r$ follows. Thus $\beta_r$ is a root of $(z^r - 2z + 1) = (z - 1)(z^{r-1} + \cdots + z - 1)$. Since $\beta_r$ is a positive real and $\beta_r < 1$ it is a positive root of $z^{r-1} + \cdots + z - 1$. This polynomial is monotone increasing for $z > 0$, thus it has only one positive root which is easily seen to be between $1/2$ and $1/2 + 1/2^r$. ∎

Note that Proposition 7.3 (i) holds for $r = 2$ as well, as $\alpha_2 = 1$.

*Theorem 7.4 ([F5])* There exists an absolute constant $c$ so that for all $t \geqslant 1$ and $r \geqslant 3$ one has

(6.4) $$\alpha_r^t \geqslant p(r, t) > c\alpha_r^t/t .$$

*Proof.* The first part follows directly from Propositions 7.2 and 7.3. To prove the second consider the family $W(n, i)$ consisting of those subsets $F$ of $X$ for which the corresponding walk $w(F)$ is going above the line $y = t + (r - 1)x$ at $x = i$. That is $W(n, i) = \{F \subseteq X : |F \cap [1, t + ri]| \geqslant t + (r - 1)i\}$. This family is clearly $r$-wise $t$-intersecting and with the previous notation $|W(n, i)| 2^{-n} \geqslant p_i$ holds for

$n \geqslant t + ri$. Direct calculation shows that for $i > c_r t$ the ratio $|\bar{W}(n, i)|2^{-n}$ decreases exponentially fast as a function of $i$, where $c_r \to 0$ as $r \to \infty$. This shows that the maximal value of $|W(n, i)|2^{-n}$ is greater than $\frac{c}{t}\Sigma p_i = c\alpha_r^t/t$ for some absolute constant $c$. ■

*Conjecture 7.5. ([F6])*

(6.5) $$f(n, r, t) = \max_i |W(n, i)| .$$

*Remark 7.6.* In [F6] the above methods were used to prove this conjecture for $t \leqslant r2^r/150$. In the case $r = 2$ the conjecture is implied by the Katona Theorem (Theorem 5.1). Simple computation shows that on the RHS of (7.5) for $t \leqslant 2^r - r - 1$ is the family $W(n, 0) = \{F \subseteq X: [1, t] \subseteq F\}$ is maximal and in fact it is the only maximal one for $t < 2^r - r - 1$. It is tied with $W(n, 1)$ if $t = 2^r - r - 1$, while for $t > 2^r - r - 1$ one has $|W(n, 1)| > |W(n, 0)|$.

Let us give the proof of the following, relatively simple case.

*Proposition 7.7.* Suppose that $t < (\ln 2)2^{r-1} - 1$ then $f(n, r, t) = 2^{n-t}$ and $W(n, 0)$ is the only optimal family.

*Proof.* Suppose $\mathcal{F} \subset 2^X$ is $r$-wise $t$-intersecting. If for some $F_1, \ldots, F_{r-1} \in \mathcal{F}$ one has $|F_1 \cap \cdots \cap F_{r-1}| = t$, then necessarily this $t$-subset of $X$ is contained in all members of $\mathcal{F}$ proving the statement. Thus we may assume that $\mathcal{F}$ is $(r-1)$-wise $(t+1)$-intersecting. Since $p(r, t)2^n \geqslant f(n, r, t)$, Theorem 7.4 implies

$$f(n, r-1, t+1) \leqslant 2^n(\alpha_{r-1})^{t+1} < 2^n \left(\frac{1}{2}\left(1 + \frac{1}{2^{r-1}}\right)\right)^{t+1} < 2^{n-t-1}e^{(t+1)/2^{r-1}}$$

which is smaller than $2^{n-t}$ for $t + 1 \leqslant 2^{r-1} \ln 2$. This yields $|\mathcal{F}| < 2^{n-1}$. ■

## 8. CROSS-INTERSECTING FAMILIES

Let $\mathcal{F}_1, \ldots, \mathcal{F}_r$ be families of subsets of $X$. We say that they are *cross-wise t-intersecting* if for all choices of $F_j \in \mathcal{F}_j$, $1 \leqslant j \leqslant r$, $|F_1 \cap \cdots \cap F_r| \geqslant t$ holds.

It is easy to check that if we apply the $(i, j)$-shift simultaneously to $\mathcal{F}_1, \ldots, \mathcal{F}_r$ then the resulting families $S_{ij}(\mathcal{F}_1), \ldots, S_{ij}(\mathcal{F}_r)$ will be cross-wise $t$-intersecting.

Thus when we are interested in bounds on $f(|\mathcal{F}_1|, \ldots, |\mathcal{F}_r|)$, we may suppose that $\mathcal{F}_1, \ldots, \mathcal{F}_r$ are stable. The next proposition is a sharpening of Proposition 7.2.

*Proposition 8.1.* Suppose that $\mathcal{F}_1, \ldots, \mathcal{F}_r$ are stable, cross-wise $t$-intersecting families, $F_j \in \mathcal{F}_j$ is arbitrary but fixed, $1 \leqslant j \leqslant r$. Then there exists $\ell \geqslant t$ such that

(8.1) $$\sum_{1 \leqslant j \leqslant r} |F_j \cap [1, \ell]| \geqslant (r-1)\ell + t \text{ holds.}$$

Note that (8.1) is equivalent to $\sum |[1, \ell] - F_j| \leqslant \ell - t$ and therefore implies $|[1, \ell] \cap F_1 \cap \cdots \cap F_r| \geqslant t$.

*Proof of (8.1).* Suppose that (8.1) does not hold for some $F_1 \in \mathscr{F}_1, \ldots, F_r \in \mathscr{F}_r$ and among such $F_j$ suppose that $F_1, \ldots, F_r$ is chosen so that $|F_1 \cap \cdots \cap F_r|$ is minimal.

Choose $\ell$ as the minimal integer satisfying

$$(8.2) \qquad |F_1 \cap \cdots \cap F_r \cap [1, \ell]| = t$$

If there exists $1 \leqslant i < \ell$ such that $i$ is not contained in at least two out of $F_1, \ldots, F_r$ then choose $1 \leqslant p < q \leqslant r$ with $i \notin F_p$, $i \notin F_q$ and define $\tilde{F}_p = (F_p - \{\ell\}) \cup \{i\}$, $\tilde{F}_s = F_s$ for $s \neq p$. Since $\mathscr{F}_p$ is stable, $\tilde{F}_p \in \mathscr{F}_p$ and

$$(8.3) \qquad \tilde{F}_1 \cap \cdots \cap \tilde{F}_r = F_1 \cap \cdots \cap F_r - \{\ell\} \text{ holds.}$$

By minimality, there exists $\tilde{\ell} \geqslant 0$, such that

$$(8.4) \qquad \sum_{1 \leqslant j \leqslant r} |\tilde{F}_j \cap [1, \tilde{\ell}]| \geqslant (r-1)\tilde{\ell} + t > \sum_{1 \leqslant j \leqslant r} |F_j \cap [1, \tilde{\ell}]| .$$

Comparing the extreme sides of (8.4) yields $\tilde{\ell} < \ell$. Consequently,

$$|\tilde{F}_1 \cap \cdots \cap \tilde{F}_r \cap [1, \ell]| \geqslant |\tilde{F}_1 \cap \cdots \cap \tilde{F}_r \cap [1, \tilde{\ell}]| \geqslant t .$$

This, however contradicts (8.2) and (8.3). ■

Hujter observed that Proposition 8.1 has the following surprising corollary.

*Proposition 8.2. (Hujter [Hu])* Suppose that $\mathscr{F}_1, \ldots, \mathscr{F}_r \subset 2^X$ are stable, cross-wise $t$-intersecting. Let $1 \leqslant j \leqslant r$ and let $F_j$, $G_j$ be arbitrary sets satisfying $F_j - [1, t] = G_j - [1, t]$, $|F_j| = |G_j|$ and $F_j \in \mathscr{F}_j$. Then adding $G_j$ to $\mathscr{F}_j$ will not destroy the cross-wise $t$-intersecting property.

*Proof.* Let $F_s \in \mathscr{F}_s$ be arbitrary, $1 \leqslant s \neq j \leqslant r$. Since $|F_j \cap [1, \ell]| = |G_j \cap [1, \ell]|$ for all $\ell \geqslant t$, Proposition 8.1 implies the existence of $\ell \geqslant t$ with

$$|G_j \cap [1, \ell]| + \sum_{s \neq j} |F_s \cap [1, \ell]| \geqslant (r-1)\ell + t \text{ and thus}$$

$$|G_j \cap (\bigcap_{s \neq j} F_s) \cap [1, \ell]| \geqslant t \ ■$$

For a family $\mathscr{F} \subset 2^X$ define $\lambda(\mathscr{F}) = \max\{\ell : \forall F \in \mathscr{F}, \exists i \geqslant 0$ with $|F \cap [1, ri + \ell]| \geqslant (r-1)i + \ell\}$. Clearly, $\lambda(\mathscr{F}) \geqslant 0$. In the geometric language, $\ell$ is the largest integer such that no walk $w(F)$, $F \in \mathscr{F}$, lies entirely under the line $y = (r-1)x + \ell$.

The next proposition extends Proposition 7.2 in another way.

*Proposition 8.3.* Suppose that $\mathcal{F}_1, \dots, \mathcal{F}_r \subset 2^X$ are stable and cross-wise $t$-intersecting. Then

(8.5) $$\lambda(\mathcal{F}_1) + \cdots + \lambda(\mathcal{F}_r) \geqslant rt \text{ holds .}$$

*Proof.* Set $\lambda_j = \lambda(\mathcal{F}_j)$. By stability for $1 \leqslant j \leqslant r$ we can choose $F_j \in \mathcal{F}_j$ satisfying

$$F_j \subset \{1, 2, \dots, \lambda_j, \lambda_j + 2, \lambda_j + 3, \dots, \lambda_j + r, \lambda_j + r + 2, \lambda_j + r + 3, \dots\} \text{ i.e.,}$$

$$F_j \subset [1, n] - \{\lambda_j + 1 + ir : i \geqslant 0\} .$$

Note that for every $\ell \geqslant 0$

(8.6) $$|[1, \ell] - F_j| \geqslant \lceil (\ell - \lambda_j)/r \rceil \text{ holds .}$$

In view of Proposition 8.1 we can choose $\ell \geqslant 0$ such that

(8.7) $$\sum_j |[1, \ell] - F_j| \leqslant \ell - t$$

Summing (8.6) for $1 \leqslant j \leqslant r$ and using (8.7) gives

$$\ell - t \geqslant \sum_j |[1, \ell] - F_j| \geqslant \sum_{1 \leqslant j \leqslant r} \lceil (\ell - \lambda_j)/r \rceil \geqslant \ell - \sum_j \lambda_j / r .$$

Comparing the extreme sides (8.5) follows. ■

Recall the definition of $\alpha_r$ from Proposition 7.3.

*Corollary 8.4* Suppose that $\mathcal{F}_1, \dots, \mathcal{F}_r \subset 2^X$ are cross-wise $t$-intersecting. Then

(8.8) $$|\mathcal{F}_1| \cdot \dots \cdot |\mathcal{F}_r| < 2^{nr} \alpha_r^{tr} .$$

*Proof.* In proving (8.8) we may assume that $\mathcal{F}_j$ is stable. Then Theorem 7.4 implies $|\mathcal{F}_j| < 2^n \alpha_r^{\lambda(\mathcal{F}_j)}$. Taking products over $1 \leqslant j \leqslant r$ and using (8.5) the inequality (8.8) follows.

## 9. SOME NUMERICAL EXAMPLES

In this section first we will bound the maximum size $f(n, 3, t)$ of 3-wise $t$-intersecting families for $t \leqslant 6$. In particular, we will show that

$$f(n, 3, t) = 2^{n-t} \text{ for } t \leqslant 3 .$$

Throughout this section $\mathcal{F}$ is a 3-wise $t$-intersecting stable family on $[1,n] = \{1, 2, \dots, n\}$.

For sets $A \subset B$ let us set $\mathcal{F}(A, B) = \{F - B : F \cap B = A, F \in \mathcal{F}\}$.

*Claim 9.1.* Either $\cap \mathcal{F} \supset [1,t]$ and thus $|\mathcal{F}| \leqslant 2^{n-t}$ or

(9.1) $$|\mathcal{F}([1,t], [1,t])| \leqslant 2^{n-t-1}$$

*Proof.* If (9.1) does not hold, then by Theorem 1.0 we can find two sets $F, F' \in \mathcal{F}$ with $F \cap F' = [1,t]$ implying $[1,t] \subset F''$ for all $F'' \in \mathcal{F}$. ■

Recall the definition of $\alpha_3$ from Proposition 7.3 and note that $\alpha_3 = (\sqrt{5} - 1)/2$. Propositions 7.2 and 7.3 imply $f(n, 3, t) < 2^n \alpha_3^t$. The next proposition gives a slight improvement.

*Claim 9.2.* For $t \geqslant 2$

(9.2) $$f(n, 3, t) \leqslant 2^n (\alpha_3^t - 2^{-t-1}) \text{ holds}.$$

*Proof.* Since $\alpha_3^t - 2^{-t-1} > 2^{-t}$ for $t \geqslant 2$, we may suppose that $|\cap \mathcal{F}| \leqslant t - 1$. By Proposition 7.2 every walk $w(F)$ with $F \in \mathcal{F}$ hits the line $y = 2x + t$. By Proposition 7.3 there are less than $\alpha_3^t 2^n$ subsets $F \in 2^{[1,n]}$ such that $w(F)$ hits this line. Among these sets $2^{n-t}$ contain $[1,t]$. But Claim 9.1 implies that at least $2^{n-t-1}$ out of these sets are not in $\mathcal{F}$. ■

*Proposition 9.3.* For $s \leqslant t$ and $A \subset [1, s]$ the family $\mathcal{F}(A, [1, s])$ is 3-wise $(t + 2s - 3|A|)$-intersecting.

*Proof.* Let $F_1$, $F_2$, $F_3$ be arbitrary members of $\mathcal{F}$ with $F_i \cap [1, s] = A$, $i = 1, 2, 3$. Choose $\ell$ from Proposition 8.1 and set $G_i = F_i \cap [s + 1, \ell]$.

Then we infer

$$|G_1| + |G_2| + |G_3| = |F_1 \cap [1, \ell]| + |F_2 \cap [1, \ell]| + |F_3 \cap [1, \ell]| - 3|A|$$
$$\geqslant 2(\ell - s) + t + 2s - 3|A|,$$

yielding $|G_1 \cap G_2 \cap G_3| \geqslant t + 2s - 3|A|$. ■

*Theorem 9.4.* The following equality and inequality hold.

(9.3) $$f(n, 3, t) = 2^{n-t} \text{ for } i \leqslant t \leqslant 3,\ n \geqslant t.$$

(9.4) $$f(n, 3, 6) < 0.03149 \cdot 2^n.$$

Moreover, if $\mathcal{F}$ is 3-wise $t$-intersecting with $|\cap \mathcal{F}| < t$ then $|\mathcal{F}| < 2^{n-t}$ holds for $1 \leqslant t \leqslant 3$.

*Proof.* We prove all the upper bounds together, using induction on $n$. For $n \leqslant t$ all bounds are trivially true. Without loss of generality we may assume that $|\cap \mathcal{F}| < t$ and therefore we can apply (9.1).

Consider first the case $t = 3$. For $A \subset [1, 3]$ set $f(A) = |\mathcal{F}(A, [1, 3])|$. Clearly, we have

(9.5) $$|\mathcal{F}| = \sum_{A \subset [1,3]} f(A)$$

In view of Claim 9.1 we have

(9.6) $$f([1, 3]) \leqslant 2^{n-4}.$$

For $A \subset [1,3]$, $\mathcal{F}(A,[1,3])$ is 3-wise $(9-3|A|)$-intersecting by Proposition 9.3. The induction hypothesis yields

(9.7) $$f(A) \leqslant 2^{n-6} \text{ for all } A \subset [1,3],\ |A| = 2 \text{ and}$$

(9.8) $$f(A) \leqslant 0.03149 \cdot 2^{n-3} \text{ for all } A \subset [1,3],\ |A| = 1\,.$$

For $A = \varnothing$, (9.2) implies

(9.9) $$f(\varnothing) \leqslant 0.01218 \cdot 2^{n-3}$$

Summing (9.6), (9.7), (9.8) and (9.9) it follows from (9.5) that $|\mathcal{F}| < 0.98165 \cdot 2^{n-3}$, as desired.
Now, $f(n,3,t) \geqslant f(n-1,3,t-1)$ implies (9.3) for $t = 1, 2$ as well.
To prove (9.4) we set $f(A) = |\mathcal{F}(A,[1,6])|$ for $A \subset [1,6]$. Then

(9.10) $$|\mathcal{F}| = \sum_{A \subset [1,6]} f(A) \text{ holds}\,.$$

The next six inequalities follow from the induction hypothesis or from (9.2), using Proposition 9.3.

$$f([1,6]) \leqslant 2^{n-7}$$

$$f(A) \leqslant 2^{n-3} \qquad \text{for } A \subset [1,6],\ |A| = 5$$

$$f(A) \leqslant 0.03149 \cdot 2^{n-6} \qquad \text{for } A \in \binom{[1,6]}{4}$$

$$f(A) < 0.01218 \cdot 2^{n-6} \qquad \text{for } A \in \binom{[1,6]}{3}$$

$$f(A) < 0.00299 \cdot 2^{n-6} \qquad \text{for } A \in \binom{[1,6]}{2}$$

$$f(A) < 0.00072 \cdot 2^{n-6} \qquad \text{for } A \in \binom{[1,6]}{1} \text{ and}$$

$$f(\varnothing) < 0.00018 \cdot 2^{n-6}$$

Summing these inequalities yields in view of (9.10)

$$|\mathcal{F}| < 0.03149 \cdot 2^{n}, \text{ as desired}\,. \blacksquare$$

*Remark 9.5.* From the proof it is clear that if $\mathcal{F}$ is 3-wise 3-intersecting with $|\cap \mathcal{F}| < 3$ then actually, $|\mathcal{F}| < 0.98165 \cdot 2^{n-3}$ holds. With similar considerations one can show that if $\mathcal{F}$ is 3-wise 2-intersecting with $|\cap \mathcal{F}| < 2$ then $|\mathcal{F}| < 0.81 \cdot 2^{n-2}$. The same approach yields:

(9.11) $\qquad f(n, 5) \leqslant 0.79 \cdot 2^{n-4}$, which we will use in Section 12 .

## 10. PAIRWISE DISJOINT SETS

We start this section by a theorem of Erdös and Gallai on 2-uniform hypergraphs, that is ordinary graphs.

*Theorem 10.1 ([EG])* Let $\mathscr{G} \subset \binom{X}{2}$ be a graph on $n$ vertices, $s \geqslant 2$, $n \geqslant 2s$ and suppose that $\mathscr{G}$ does not contain $s$ pairwise disjoint edges. Then

$$(10.1) \qquad |\mathscr{G}| \leqslant \max\left\{\binom{2s-1}{2}, \binom{s-1}{2} + (s-1)(n-s)\right\}$$

Moreover, equality holds in (10.1) if and only if either $\mathscr{G} = \binom{Y}{2}$ for some $(2s-1)$-element set $Y$ or $\mathscr{G} = \left\{G \in \binom{X}{2} : G \cap Z \neq \varnothing\right\}$ for some $Z \in \binom{X}{s-1}$.

*Proof (Akiyama-Frankl [AF]).* In proving (10.1) we may suppose again that $\mathscr{G}$ is stable. Since $\mathscr{G}$ contains no $s$ pairwise disjoint edges, one of the following $s$ subsets is not in $\mathscr{G}$.

$$G_i = \{i,\ 2s + 1 - i\},\quad i = 1, \ldots, s\ .$$

However, if $G_i \notin \mathscr{G}$, then the stability of $\mathscr{G}$ implies

$$\mathscr{G} \subset \mathscr{G}_i = \{G \in \binom{X}{2} : G \cap [1, i-1] \neq \varnothing \text{ or } G \subset [1, 2s-i]\}$$

Note that actually $G_i$ contains no $s$ pairwise disjoint edges. Now (10.1) follows from

$$\max_{1 \leqslant i \leqslant s} |\mathscr{G}_i| = |\mathscr{G}_1| \text{ or } |\mathscr{G}_s|$$

and equality holds only if

$$\mathscr{G} = \mathscr{G}_1 \text{ or } \mathscr{G} = \mathscr{G}_s\ .$$

Finally, note that if $\mathscr{G}$ contains no $s$ pairwise disjoint edges and $S_{ij}(\mathscr{G})$ is isomorphic to $\mathscr{G}_\ell$ for some $1 \leqslant \ell \leqslant s$, then $\mathscr{G}$ is isomorphic to $\mathscr{G}_\ell$ as well. ■

The families corresponding to $\mathscr{G}_1$ and $\mathscr{G}_s$ for $k$-graphs with $k \geqslant 3$ are:

$$\mathscr{F}_1 = \left\{F \in \binom{X}{k} : F \cap [1, s-1] \neq \varnothing\right\} \text{ and } \mathscr{F}_k = \binom{[1, ks-1]}{k}.$$

*Conjecture 10.2 (Erdös [E])* Suppose that $\mathscr{F} \subset \binom{X}{k}$, $n \geqslant ks$ and $\mathscr{F}$ contains no $s$ pairwise disjoint sets. Then

$$(10.2) \qquad |\mathscr{F}| \leqslant \max\left\{\binom{n}{k} - \binom{n-s+1}{k}, \binom{ks-1}{k}\right\} \text{ holds .}$$

Erdös [E] proved this conjecture for $n > n_0(k, s)$. The bounds on $n_0(k, s)$ were improved by Bollobás, Daykin and Erdös [BDS] who showed that (10.2) holds for $n > 2k^3s$. Füredi and the author (unpublished) proved (10.2) for $n > 100ks^2$, but to prove (10.2) in full generality appears to be a very difficult problem.

Let us prove an upper bound, which is not too far from (10.2) and holds for all $n \geqslant ks$.

*Theorem 10.3.* Suppose that $\mathscr{F} \subset \binom{X}{k}$, $n \geqslant ks$ and $\mathscr{F}$ contains no $s$ pairwise disjoint edges. Then

$$(10.3) \qquad |\mathscr{F}| \leqslant (s-1)\binom{n-1}{k-1} \text{ holds .}$$

*Proof.* Note that for $s = 2$, (10.3) reduces to the Erdös-Ko-Rado Theorem. In fact, our proof will be similar to that. First we prove (10.3) for $n = ks$. Let $X = G_1 \cup G_2 \cup \cdots \cup G_s$ be an arbitrary partition with $|G_1| = \cdots = |G_s| = k$. Out of these $s$ sets at most $s - 1$ can be in $\mathscr{F}$. Averaging over all partitions gives

$$|\mathscr{F}| \leqslant \frac{s-1}{s}\binom{ks}{k} = \binom{ks-1}{k}.$$

Now we apply induction on $n$ and prove the statement simultaneously for all $k$ with $ks \leqslant n$.

Again, we may assume that $\mathscr{F}$ is stable. Consider $\mathscr{F}(\bar{n}) = \{F \in \mathscr{F}: n \notin F\}$ and $\mathscr{F}(n) = \{F - \{n\}: n \in F \in \mathscr{F}\}$. We claim that neither of them contains $s$ pairwise disjoint sets. Indeed, this is trivial for $\mathscr{F}(\bar{n}) \subset \mathscr{F}$. As to $\mathscr{F}(n)$, note that if $H_1, \ldots, H_s \in \mathscr{F}(n)$ are pairwise disjoint then choosing $s$ distinct elements $y_1, \ldots, y_s$ from $[1,n] - (H_1 \cup \cdots \cup H_s)$, which has size $n - s(k-1) \geqslant s$, the stability of $\mathscr{F}$ implies $F_i = (H_i \cup \{y_i\}) \in \mathscr{F}$. However, $F_1, \ldots, F_s$ are pairwise disjoint, a contradiction.

Now using the induction hypothesis we infer that

$$|\mathscr{F}| = |\mathscr{F}(\bar{n})| + |\mathscr{F}(n)| \leqslant (s-1)\binom{n-2}{k-1} + (s-1)\binom{n-2}{k-2} = (s-1)\binom{n-1}{k-1}. \quad \blacksquare$$

## 11. ON $r$-WISE INTERSECTING FAMILIES

Recall that $\mathcal{F} \subset 2^X$ is called $r$-wise intersecting if $F_1 \cap \cdots \cap F_r \neq \emptyset$ holds for all $F_1, \ldots, F_r \in \mathcal{F}$. If $|F_1| + \cdots + |F_r| > (r-1)n$, then necessarily $F_1 \cap \cdots \cap F_r \neq \emptyset$ holds. This shows that the assumptions of the next result are necessary.

*Theorem 11.1* *([F8])* Suppose that $\mathcal{F} \subset \binom{X}{k}$ is $r$-wise intersecting, $rk \leqslant (r-1)n$. Then

$$|\mathcal{F}| \leqslant \binom{n-1}{k-1}. \tag{11.1}$$

Moreover, excepting the case $r = 2$, $n = 2k$ equality holds if and only if $\mathcal{F} = \{F \in \binom{X}{k} : x \in X\}$ holds for some $x \in X$.

Neither the original nor the present proof uses shifting. However, the present proof uses the Kruskal-Katona Theorem, which we proved by shifting.

First we prove a proposition which is due to Kleitman.

*Proposition 11.2* *([Kl])* Suppose that $\mathcal{F}_i \subset \binom{X}{k_i}$, $i = 1, \ldots, r$, $k_1 + \cdots + k_r = n$. If there are no $F_i \in \mathcal{F}_i$, $1 \leqslant i \leqslant r$ with $F_1 \cup \cdots \cup F_r = X$, then

$$\sum_{1 \leqslant i \leqslant r} |\mathcal{F}_i| / \binom{n}{k_i} \leqslant r - 1 \text{ holds}. \tag{11.2}$$

Moreover, equality holds if and only if for every ordered partition $X = G_1 \cup \cdots \cup G_r$ satisfying $|G_i| = k_i$ there is exactly one $i$, $1 \leqslant i \leqslant r$ with $G_i \notin \mathcal{F}_i$.

*Proof.* Consider all ordered partitions $X = G_1 \cup \cdots \cup G_r$ with $|G_i| = k_i$, $1 \leqslant i \leqslant r$. For a fixed $F \in \binom{X}{k_i}$ one has $F = G_i$ for a fraction $1/\binom{n}{k_i}$ of all these partitions. Thus $G_i \in \mathcal{F}_i$ holds for a fraction $|\mathcal{F}_i|/\binom{n}{k_i}$ of them. Since $G_i \notin \mathcal{F}_i$ must hold for at least one $i$, the statement follows. ■

*Proof of Theorem 11.1.* Set $\mathcal{F}^c = \{X - F : F \in \mathcal{F}\}$. Choose numbers $k_1, \ldots, k_r$ satisfying $0 \leqslant k_i \leqslant n - k$. $k_1 + \cdots + k_r = n$.

Set $\mathcal{F}_i = \partial_{k-k_i}(\mathcal{F}) = \left\{G \in \binom{X}{k_i} : \exists F \in \mathcal{F}^c, G \subset F\right\}$.

Note that the fact that $\mathcal{F}$ is $r$-wise intersecting is equivalent to $\mathcal{F}^c$ not containing $r$ sets whose union is $X$. Thus $\mathcal{F}_1, \dots, \mathcal{F}_r$ satisfy the assumptions of Proposition 11.2.

Suppose $|\mathcal{F}| \geqslant \binom{n-1}{k-1} = \binom{n-1}{n-k}$. Then, by a consequence of the Kruskal-Katona Theorem (Corollary 3.4), one has $|\mathcal{F}_i| \geqslant \binom{n-1}{k_i}$, that is $|\mathcal{F}_i| / \binom{n}{k_i} \geqslant 1 - \frac{k_i}{n}$ for $1 \leqslant i \leqslant r$. Comparing with (11.2) gives that equality must hold for all $i$. Again by Corollary 3.4 if $k_i < n - k$ for some $i$, we infer $\mathcal{F}^c = \binom{X - \{x\}}{n-k}$ for some $x \in X$.

If $k_1 = \cdots = k_r = n - k$, and thus $(r-1)n = rk$, i.e., if $\mathcal{F}_1 = \cdots = \mathcal{F}_r = \mathcal{F}^c$, then by Proposition 11.2 $|\mathcal{F}^c| \leqslant \frac{r-1}{r}\binom{r(n-k)}{n-k} = \binom{n-1}{k-1}$ with equality only if there is one missing set from every partition. This implies that $\binom{X}{n-k} - \mathcal{F}^c$ is an intersecting family of size $\binom{n-1}{n-k-1}$.

Thus, for $r \geqslant 3$, that is, for $(n-k) \leqslant n/3$, the uniqueness of the optimal families in the Erdös-Ko-Rado Theorem implies $\binom{X}{n-k} - \mathcal{F}^c = \left\{G \in \binom{X}{n-k} : x \in G\right\}$ for some $x \in X$, and the uniqueness part of Theorem 11.1 follows. ■

## 12. A Helly-type theorem

From various previous theorems we know that if $\mathcal{F} \subset 2^X$ is $r$-wise intersecting with $|\mathcal{F}| \geqslant 2^{n-1}$, then for $r \geqslant 3$ $\mathcal{F}$ consists of all subsets through some fixed element. What if we bar this family? Define

$$\mathcal{H} = \mathcal{H}(n,r) = \{H \subset X: |H \cap [1,r+1]| \geqslant r\}.$$

Clearly, $\mathcal{H}$ is $r$-wise intersecting, $\cap\, \mathcal{H} = \emptyset$ and $|\mathcal{H}| = (r+2)2^{n-r-1}$ hold.

*Theorem 12.1 (Brace and Daykin [BD]).* Suppose that $\mathcal{F} \subset 2^X$ is $r$-wise intersecting, $\cap\, \mathcal{F} = \emptyset$ then $|\mathcal{F}| \leqslant |\mathcal{H}(n,r)|$; moreover, for $r \geqslant 3$ equality holds if and only if $\mathcal{F}$ is isomorphic to $\mathcal{H}(n,r)$.

The original proof of this powerful result did not use shifting. Kleitman (cf. [P]) gave a proof using shifting. Here we present an alternate proof which is based on the following.

*Proposition 12.2.* Suppose that $\mathcal{G} \subset 2^X$ is $r$-wise $r$-intersecting, $r \geqslant 3$. Then $|\mathcal{G}| \leqslant 2^{n-r}$ with equality holding if and only if $\mathcal{G} = \{G \subset X: T \subset G\}$ for some $T \in \binom{X}{r}$.

*Proof.* For $r = 3$ this result is contained in Theorem 9.4. For $r \geqslant 6$, the statement follows from Proposition 7.7. Suppose now that $r = 4$ or 5. If $\mathcal{G}$ is not $(r-1)$-wise $(r+1)$-intersecting, then we can find an $r$-element set $T$ and $G_1, \ldots, G_{r-1} \in \mathcal{G}$ with $G_1 \cap \cdots \cap G_{r-1} = T$. Consequently, $T \subset G$ for every $G \in \mathcal{G}$.

Suppose next that $\mathcal{G}$ is $(r-1)$-wise $(r+1)$-intersecting. If $r = 4$, then (9.11) implies $|\mathcal{G}| < 0.79 \cdot 2^{n-4}$, and we are done. If $r = 5$, then Propositions 7.2 and 7.3 imply $|\mathcal{G}| < 2^n \alpha_4^6$. One can check that $\alpha_4 < 0.544$ and thus $|\mathcal{G}| < 2^{n-5}$ concluding the proof. ■

*Proof of Theorem 12.1.* We may assume that $\mathcal{G}$ is a filter, i.e., $G \subset H \subset X$ and $G \in \mathcal{G}$ imply $H \in \mathcal{G}$. Since $\cap\,\mathcal{G} = \emptyset$, $X - \{i\}$ is in $\mathcal{G}$ for all $1 \leqslant i \leqslant n$. As this is maintained by shifting, we may assume that $\mathcal{G}$ is stable. For $r = 2$ one has $(r+2)2^{n-r-1} = 2^{n-1}$. Thus the statement is trivially true. Apply induction and suppose that for $r-1$ the theorem is proved. Consider the families $\mathcal{G}(1)$ and $\mathcal{G}(\bar{1})$. Since $\mathcal{G}$ is $r$-wise intersecting, $\mathcal{G}(1)$ is $(r-1)$-wise intersecting on $X - \{1\}$ and $(X - \{i\}) \in \mathcal{G}$ for $2 \leqslant i \leqslant n$ implies $\cap\,\mathcal{G}(1) = \emptyset$. By the induction hypothesis we infer:

$$|\mathcal{G}(1)| \leqslant (r+1)2^{n-r-1}\,. \tag{12.1}$$

From Proposition 8.1 it follows, in the same way as Proposition 9.3, that $\mathcal{G}(\bar{1})$ is $r$-wise $r$-intersecting on $X - \{1\}$. Thus Proposition 12.2 implies:

$$|\mathcal{G}(\bar{1})| \leqslant 2^{n-r-1}\,. \tag{12.2}$$

Adding (12.1) and (12.2) we obtain $|\mathcal{G}| \leqslant (r+2)2^{n-r-1}$, as desired.

In case of equality, equality must hold in (12.2). Consequently, $[2,r+1] \in \mathcal{G}$, which implies $|H \cap [2,r+1]| \geqslant r-1$ for $H \in \mathcal{G}(1)$.

We infer $|G \cap [1,r+1]| \geqslant r$ for all $G \in \mathcal{G}$, i.e., $\mathcal{G} \subset \mathcal{H}(n,r)$, as desired. ■

*Remark 12.3.* Let us note that the present proof shows (via Remark 9.5) that if $\mathcal{G} \neq \mathcal{H}(n,r)$ then $|\mathcal{G}| < (r+1.982)2^{n-r-1}$.

## REFERENCES


[A] N. Alon, Ph.D. Thesis, Hebrew University, Jerusalem, 1983.

[AF] J. Akiyama and P. Frankl, On the size of graphs with complete factors, J. Graph Th. 9 (1985) 197-201.

[BD] A. Brace and D. E. Daykin, A finite set covering theorem, Bull. Austral. Math. Soc. 5 (1971) 197-202.

[BDE] B. Bollobás, D. E. Daykin and P. Erdös, Sets of independent edges of a hypergraph, Quart. J. Math. Oxford 21 (1976) 25-32.

[DF] D. E. Daykin and P. Frankl, Sets of finite sets satisfying union conditions, Mathematika 29 (1982) 128-134.

[E] P. Erdoš, A problem of independent $r$-tuples, Annales Univ. Budapest, 8 (1965) 93-95.

[EG] P. Erdös and T. Gallai, On the maximal paths and circuits of graphs, Acta Math. Hungar. 10 (1959) 337-357.

[EKR] P. Erdös, Chao Ko and R. Rado, Intersection theorems for systems of finite sets, Quart. J. Math. Oxford 12 (1961) 313-320.

[F1] P. Frankl, The Erdös-Ko-Rado Theorem is true for $n = ckt$, Coll. Soc. Math. J. Bolyai 18 (1978) 365-375.

[F2] P. Frankl, A new short proof for the Kruskal-Katona theorem, Discrete Math. 48 (1984) 327-329.

[F3] P. Frankl, New proofs for old theorems in extremal set theory, Proc. Coll. ISI, Calcutta (1985) 127-132.

[F4] P. Frankl, Erdös-Ko-Rado Theorem with conditions on the maximal degree, J. Combinatorial Th. A., submitted.

[F5] P. Frankl, Ph.D. Thesis, Eötvös University, Budapest, 1976.

[F6] P. Frankl, Families of finite sets satisfying a union condition, Discrete Math. 26 (1979) 111-118.

[F7] P. Frankl, Families of finite sets satisfying an intersection condition, Bull. Austral. Math. Soc. 15 (1976) 73-79.

[F8] P. Frankl, On Sperner families satisfying an additional condition, J. Combinatorial Th. A. 20 (1976) 1-11.

[FF1] P. Frankl and Z. Füredi, Non-trivial intersecting families, J. Combinatorial Th. A. 41 (1986) 150-153.

[FF2] P. Frankl and Z. Füredi, More on the Erdös-Ko-Rado Theorem: beyond stars, manuscript.

[FG] Z. Füredi and J. R. Griggs, Families of finite sets having minimum shadows, Combinatorica 6 (1986) 393-401.

[HM] A. J. W. Hilton and E. C. Milner, Some intersection theorems for systems of finite sets, Quart. J. Math. Oxford 18 (1967) 369-384.

[Hu] M. Hujter, personal communication, 1981.

[Ka1] G. O. H. Katona, A theorem on finite sets, in Theory of Graphs, Proc. Colloq. Tihany, Budapest, 1968, 187-207.

[Ka2] G. O. H. Katona, Intersection theorems for systems of finite sets, Acta Math. Hungar. 15 (1964) 329-337.

[Kl] D. J. Kleitman, Maximum number of subsets of a finite set no $k$ of which are pairwise disjoint, J. Combinatorial Th. 5 (1968) 157-163.

[Kr] J. B. Kruskal, The number of simplices in a complex, in Math. Optimization Techniques, Univ. California Press, Berkeley, 1963, 251-278.

[L] L. Lovász, Problem 13.31, Combinatorial Problems and Exercises, North Holland, Amsterdam.

[M] M. Mörs, A generalization of a theorem of Kruskal, Graphs and Comb. 1 (1985) 167-183.

[P] G. W. Peck, A Helly theorem for sets, SIAM J. Alg. Disc. Meth. 2 (1981) 306-308.

[S] E. Sperner, Ein Satz über Untermengen einer endlichen Menge, Math. Z. 27 (1928) 544-548.

[W] R. M. Wilson, The exact bound in the Erdös-Ko-Rado Theorem, Combinatorica, 4 (1984) 247-257.

## NUMBERS IN RAMSEY THEORY

*R. L. Graham*
AT&T Bell Laboratories
Murray Hill, New Jersey

and

*V. Rödl*
Czech Tech. Univ., Praha

### 1. Introduction

Ramsey theory can be loosely described as the study of structure which is preserved under finite decomposition. Its underlying philosophy is captured succinctly by the statement "Complete disorder is impossible". Since the publication of the seminal paper of Ramsey [R30] in 1930, the subject has grown with increasing vitality, and is currently among the most active areas in combinatorics. However, most of the recent work has focussed on far-ranging generalizations of the original concepts, dealing for example, with extensions to $n$-dimensional vector spaces, (and their cominatorial analogues, $n$-parameter sets), lattices, groups, various transfinite sets, induced and restricted variations, etc. (cf. [GRS80], [G83], [NR78], [NR79]). Progress on determining the basic numbers themselves, the so-called Ramsey numbers $r(k, \ell)$, has been painfully slow. We recall that for positive integers $k$ and $\ell$, $r(k, \ell)$ denotes the *least* integer $n$ such that if the edges of the complete graph $K_n$ are arbitrarily 2-colored, say with colors red and blue, then there is always formed either a $K_k$ with all edges red, or a $K_\ell$ with all edges blue. (The existence of $r(k, \ell)$ is guaranteed by Ramsey's Theorem.) For example, the first lower bound for $r(k, k)$, due to Erdös [E47], was published in 1947:

$$r(k, k) > (1 + o(1)) \frac{1}{e\sqrt{2}} \cdot k \cdot 2^{k/2}. \tag{1.1}$$

In the 40 years since its proof, this bound has only been improved by a factor of 2!

However, some breakthroughs have occurred recently, such as the first significant improvement on the upper bound

$$r(k, k) \leqslant \binom{2k-2}{k-1},$$

originally given by Erdös and Szekeres [ES35] in 1935. It is now known (cf.[R]) that for suitable positive constants $c_1$ and $c_2$,

$$r(k, k) < c_1 \binom{2k-1}{k-1} / (\log k)^{c_2}, \tag{1.2}$$

(where all logarithms in this paper are to the base $e$). We will give a proof (Theorem 2.14) that

$$r(k, k) < 6 \binom{2k-2}{k-1} / \log\log k \tag{1.3}$$

for $k$ sufficiently large.

In this paper, we will attempt to describe the current state of knowledge in this and some related areas. We will usually not give the arguments which lead to the sharpest bounds known, but rather we will concentrate on outlining some of the basic methods needed for the various improvements discussed. We will, however, try to provide appropriate references to the sharper results when we can.

## 2. Asymptotic bounds on classical problems

In this section we will review a few basic bounds on the numbers $r(k, \ell)$ and then outline some very recent improvements.

### 2.1 Lower bounds

*Theorem 2.1* [E47]

$$r(k, k) > (1 + o(1)) \frac{1}{e\sqrt{2}} k \cdot 2^{k/2} .$$

*Proof*: We show that if

$$\binom{n}{k} 2^{1-\binom{k}{2}} < 1 \tag{2.1}$$

then in fact $r(k, k) > n$, i.e., there exists a 2-coloring of $K_n$ containing no monochromatic $K_k$. (This is clearly sufficient since a simple calculation shows that if (2.1) fails then

$$n > \left(\frac{1}{e\sqrt{2}} + o(1)\right) k \cdot 2^{k/2} .$$

Consider a random 2-coloring of $K_n$, where the color of each edge is chosen independently with probability 1/2. Let $T$ be a subset of the vertex set of size $|T| = k$. The probability of the event $A_T$ that the set $[T]^2$ of pairs of $T$ is monochromatic is

$$Pr(A_T) = 2^{1-\binom{k}{2}} .$$

Thus, the probability of the event that for some $k$-element set $T$ of vertices, the set $[T]^2$ is monochromatic, is bounded above by

$$\binom{n}{k} 2^{1-\binom{k}{2}} .$$

Hence, if (2.1) holds, then there must exist a 2-coloring of the edges of $K_n$ having no monochromatic $K_k$. ■

Note that this proof, which was one of the earliest uses of the "probabilistic method", gives no information about how such a 2-coloring might actually be constructed. The best constructive lower bounds currently known (see (2.12)) are much weaker.

The bound in Theorem 2.1 has been improved slightly by Spencer [Sp75] to

$$r(k, k) > (1 + o(1)) \frac{\sqrt{2}}{e} k \cdot 2^{k/2} , \tag{2.2}$$

which is at present the sharpest known.

The proof of (2.2) is based on an ingenious probabilistic inequality of Lovász [EL75] known as the "local lemma". The setting for this result is the following. Let $\Omega$ be a probability space and let $A_1, A_2, \ldots, A_n$ be events in $\Omega$. We say that the graph $\Gamma$ with vertex set $\{1, 2, \ldots, n\}$ is a *dependence graph* of $\{A_1, A_2, \ldots, A_n\}$ if:

$$\{i\} \textit{ not } \text{joined to } J_1, J_2, \ldots, J_s \Rightarrow A_i \text{ and } A_{J_1} \cap A_{J_2} \cap \cdots \cap A_{J_s} \text{ are independent} .$$

*Theorem 2.2* (Lovász local lemma). Let $A_1, A_2, \ldots, A_n$ be events in a probability space $\Omega$ with dependence graph $\Gamma$. Suppose there exist $x_1, x_2, \ldots, x_n$ with $0 < x_i \leqslant 1$ such that

$$Pr(A_i) \leqslant (1 - x_i) \prod_{\{i,j\} \in \Gamma} x_j , \quad 1 \leqslant i \leqslant n .$$

Then

$$Pr(\bigcap_i \bar{A}_i) > 0 .$$

The proof of this version can be found in [Sp77]. A slightly more convenient form of the local lemma results from the following observation. Set

$$y_i = \frac{1 - x_i}{x_i Pr(A_i)}$$

so that

$$x_i = \frac{1}{1 + y_i Pr(A_i)} .$$

Since $1 + z \leqslant \exp(z)$, we have

*Corollary 2.3.*

Suppose $A_1, A_2, \ldots, A_n$ are events in a probability space having dependence graph $\Gamma$, and there exist positive $y_1, y_2, \ldots, y_n$ satisfying

$$\log y_i > \sum_{\{i,j\} \in \Gamma} y_j Pr(A_j) + y_i Pr(A_i)$$

for $1 \leqslant i \leqslant n$. Then

$$Pr(\bigcap_i \bar{A}_i) > 0 .$$

We next illustrate the use of this result in deriving a lower bound for $r(k, 3)$. The first bound on $r(k, 3)$, again due to Erdös [E61], was that $r(k, 3) > ck^2/(\log k)^2$ for a suitable $c > 0$. The following result gives the sharpest known bound currently known.

*Theorem 2.4* [Sp77]

$$r(k, 3) \geqslant \left[\frac{1}{27} + o(1)\right] k^2/(\log k)^2 . \tag{2.3}$$

*Proof*: The proof is a modification of that of Spencer [Sp77]. Let the edges of $K_n$ be independently 2-colored red and blue with the probability that an edge is colored red always being $p$. To each 3-element subset of vertices $S$ associate the event $A_S$ that all the edges spanned by $S$ have been colored red. Similarly, to each $k$-element subset $K$ associate the event $B_K$ that all the edges spanned by $K$ have been colored blue. Observe that

$$r(k, 3) > n \text{ if } Pr(\bigcap_S \bar{A}_S \cap \bigcap_K \bar{B}_K) > 0 .$$

Let $\Gamma$ denote the graph with $\binom{n}{3} + \binom{n}{k}$ vertices corresponding to all possible $A_S$ and $B_K$, where $\{A_S, B_K\}$ is an edge of $\Gamma$ if and only if $|S \cap K| \geqslant 2$ (i.e., the events $A_S$ and $B_K$ are dependent), and the same applies to pairs of the form $\{A_S, A_{S'}\}$ and $\{B_K, B_{K'}\}$. Let $N_{AA}$ denote the number of vertices of the form $A_S$ for some $S$ joined to some other vertex of this form (so that $N_{AA} = 3(n-2)$), and let $N_{AB}$, $N_{BA}$ and $N_{BB}$ be defined analogously. In this case, Corollary 2.3 implies:

If there exist positive $p, y, z$ such that:

$$p < 1\,,$$

(2.4)
$$\log y > yPr(A_S)(N_{AA}+1) + zPr(B_K)N_{AB}\,,$$
$$\log z > yPr(A_S)N_{BA} + zPr(B_K)(N_{BB}+1)$$

then

$$r(k,3) > n\,.$$

Now,

$$Pr(A_S) = p^3\,,\ Pr(B_K) = (1-p)^{\binom{k}{2}} \leqslant \exp\left(-p\binom{k}{2}\right).$$

Also, we have the bounds

$$N_{AB} \leqslant \binom{n}{k} < \left(\frac{ne}{k}\right)^k,\ N_{BB}+1 < \left(\frac{ne}{k}\right)^k,$$

$$N_{AA}+1 < 3n\,,\ N_{BA} = \binom{k}{2}(n-k) + \binom{k}{3} < \frac{1}{2}k^2 n\,.$$

Set

$$p = c_1 n^{-1/2}\,, \qquad k = c_2 n^{1/2}\log n$$

$$z = \exp(c_3 n^{1/2}(\log n)^2)\,,\ y = 1+\epsilon$$

where $c_1$, $c_2$, $c_3$, $\epsilon$ are positive constants to be specified later. We now verify (2.4).

First, observe that

$$zPr(B_K)\max\{N_{AB}, N_{BA}\} < zPr(B_K)\left(\frac{ne}{k}\right)^k$$

$$< \exp\left\{n^{1/2}(\log n)^2\left[c_3 + c_2\left(\frac{1-c_1c_2}{2}\right) + o(1)\right]\right\}.$$

Thus, if

(2.5)
$$c_3 + c_2\left(\frac{1-c_1c_2}{2}\right) < 0$$

then the second line of (2.4) will hold for $k$ large.

In a similar way, if in addition

(2.6)
$$(1+\epsilon)c_1^3c_2^2 < 2c_3$$

then the last line of (2.4) will hold. Finally, by choosing

$$c_1 = \frac{1}{\sqrt{3}} + o(1)\,,\ c_2 = 3\sqrt{3}/2 + o(1)\,,\ c_3 = \frac{3\sqrt{3}}{8} + o(1)\,,\ \epsilon = o(1)$$

appropriately, an easy calculation shows that (2.5) and (2.6) hold. This implies

$$k = \left(\frac{3\sqrt{3}}{2} + o(1)\right) n^{1/2} \log n$$

which in turn implies (2.3). ∎

We remark that in the same way the following more general result can be established.

*Theorem 2.5* [Sp77]

Let $\ell \geqslant 3$ be fixed. Then for a suitable constant $c = c(\ell) > 0$,

(2.7) $$r(k, \ell) > c(k/\log k)^{(\ell+1)/2}$$

The best available lower bound for general $k$ and $\ell$ comes from the following result (by using the probability method).

*Theorem 2.6* [Sp75]

If for some $p \in (0,1)$ we have

(2.8) $$\binom{n}{k} p^{\binom{k}{2}} + \binom{n}{\ell} (1-p)^{\binom{\ell}{2}} < 1$$

then $r(k,\ell) > n$.

In particular, it follows from Theorem 2.6 that

(2.9) $$r(k,\ell) > c\binom{k+\ell-2}{k-1}^{1/4}$$

for an absolute constant $c > 0$. Erdös conjectures that for fixed $\ell$,

$$r(k, \ell) > k^{\ell-1}/(\log k)^{c_\ell}$$

for a suitable constant $c_\ell > 0$ and $k$ sufficiently large.

**2.2 Constructive lower bounds**

In the preceding sections, all of the bounds given were based on the use of the probability method. As a consequence, the proofs do not produce any explicit colorings but rather, they only prove that such colorings exist. To remedy this not entirely satisfactory state of affairs, attempts have been made over the years to construct good colorings, unfortunately without much success. For the case of $r(k, 3)$, Erdös [E66] has given a construction which shows that

$$r(k, 3) > k^{\alpha + o(1)}$$

where

$$\alpha = \frac{2}{3}\left(\frac{\log 2}{\log 3 - \log 2}\right) = 1.139\ldots,$$

improving an earlier construction of his which gave a somewhat weaker result (cf. [E57]). A breakthrough for $r(k, k)$ finally occurred several years ago, however, with the result of Frankl [F77], who gave the first Ramsey construction which grew faster than any polynomial. In this section, we will outline a more recent theorem of Frankl and Wilson [FW81] on set intersections which yields the best constructive bound for $r(k, k)$ currently available.

*Theorem 2.7* [FW81]

Suppose $\mathscr{F}$ is a family of $k$-sets of $\{1, 2, \ldots, n\}$ such that for some prime power $q$,

$$F, F' \in \mathscr{F}, F \neq F' \Rightarrow |F \cap F'| \not\equiv k \pmod{q} .$$

Then

(2.10) $$|\mathscr{F}| \leqslant \binom{n}{q-1} .$$

*Proof*: Let $A_1, A_2, \ldots, A_{\binom{n}{j}}$ be all the $j$-element subsets and $B_1, B_2, \ldots, B_{\binom{n}{i}}$ be all the $i$-element subsets of $\{1, 2, \ldots, n\}$, where $i < j$. Define the $\binom{n}{i}$ by $\binom{n}{j}$ matrix $N(i, j)$ as follows: the $(u, v)$-entry of $N(i, j)$ is 1 if $B_u \subset A_v$, and 0 if $B_u \not\subset A_v$, for $1 \leqslant u \leqslant \binom{n}{i}$, $1 \leqslant v \leqslant \binom{n}{j}$. Let $V$ denote the vector space (over $\mathbf{R}$) generated by the row vectors $V_1, V_2, \ldots, V_{\binom{n}{q-1}}$ of $N(q-1, k)$. Of course

(2.11) $$\dim V \leqslant \binom{n}{q-1} .$$

It is easy to check that

$$N(i, q-1) N(q-1, k) = \binom{k-i}{q-1-i} N(i, k) .$$

Thus, for $0 \leqslant i < q-1$, the row vectors of $N(i, k)$ belong to $V$.

Consider the product

$$M(i, k) := N(i, k)^t N(i, k) .$$

The $(u, v)$-entry of the $\binom{n}{k}$ by $\binom{n}{k}$ matrix $M(i, k)$ is $\binom{|A_u \cap A_v|}{i}$ for $1 \leqslant u, v \leqslant \binom{n}{k}$. Moreover, the row vectors of $M(i, k)$ are linear combinations of the rows of $N(i, k)$, and consequently, belong to $V$.

Now, choose real numbers $a_i$, $0 \leqslant i < q$, so that

$$\sum_{i=0}^{q-1} a_i \binom{x}{i} = \binom{x-k-1}{q-1} .$$

Set $M := \sum_{i=0}^{q-1} a_i M(i, k)$ where addition is performed componentwise, i.e., the $(u, v)$-entry of $M$ is

$$m(u, v) = \sum_{i=0}^{q-1} a_1 \binom{|A_u \cap A_v|}{i} .$$

It follows from the definition that the row vectors of $M$ are in $V$, and consequently

$$\operatorname{rank} M \leqslant \dim V .$$

Let $M(\mathscr{F})$ denote the minor of $M$ spanned by the elements $m(u, v)$ for which $A_u, A_v \in \mathscr{F}$. Since for $q = p^\alpha$, $p$ prime and $\alpha \geqslant 1$, we have

$$\binom{a}{q-1} \equiv 0 \pmod{p} \textit{ iff } a \not\equiv -1 \pmod{q}$$

then it follows that

$$m(u, v) \equiv 0 \pmod{p} \text{ for } u \neq v \text{, and}$$

$$m(u, u) \not\equiv 0 \pmod{p} .$$

Thus, $\det M(\mathscr{F}) \not\equiv 0 \pmod{p}$ and so,

$$|\mathscr{F}| = \text{rank } M(\mathscr{F}) \leqslant \text{rank } M \leqslant \binom{n}{q-1}$$

as required. ■

*Corollary 2.8.*

Let $q$ be a prime power and form the graph $G = G(n, q)$ with vertex set

$$V(G) = \{F \subseteq \{1, 2, \ldots, n\} : |F| = q^2 - 1\}$$

and edge set

$$E(G) = \{\{F, F'\} : |F \cap F'| \not\equiv -1 \pmod q\}\ .$$

Then neither $G$ nor its complement $G'$ contains a complete subgraph on more than $\binom{n}{q-1}$ vertices.

*Proof*: If $F_1, F_2, \ldots, F_m$ forms a complete subgraph in $\bar{G}$ then $|F_i \cap F_j| \not\equiv -1 \pmod q$ for every $i \neq j$, and so, by Th. 2.7, we have the desired conclusion. If $F_1, F_2, \ldots, F_m$ forms a complete subgraph in $\bar{G}$ then

$$|F_i \cap F_j| \in \{q-1, 2q-1, \ldots, q^2 - q - 1\}$$

for $i \neq j$. However, we can now apply the following result of Ray-Chaudhuri and Wilson [RW75] (which can be proved along the lines of Theorem 2.7):

If $\mathscr{F}$ is a family of $k$-sets of $\{1, 2, \ldots, n\}$ and

$$s := |\{|F \cap F'| : F, F' \in \mathscr{F}, F \neq F'\}|$$

then

$$|\mathscr{F}| \leqslant \binom{n}{s}.$$

Thus,

$$m \leqslant \binom{m}{q-1}$$

in this case as well, and the Corollary is proved. ■

Choosing $n = p^3$, $q = p$, we obtain

$$r(k, k) \geqslant \exp((1 + o(1)) \log^2 k/4 \log\log k) \tag{2.12}$$

which is the best constructive lower bound on $r(k, k)$ currently known.

A graph which has often been suggested as a natural candidate for giving an exponentially large (constructive) Ramsey bound is the following. Let $p$ be a prime which is congruent to 1 modulo 4. Form the graph $G_p$ with vertex set $\mathbf{Z}_p$ (the integers modulo $p$) and edge set $\{\{i, j\} : 0 \neq i - j$ is a quadratic residue modulo $p\}$. Since $p \equiv 1 \pmod 4$, $\{i, j\}$ is an edge *iff* $\{j, i\}$ is. Recent results of Shearer [Sh86] show that for small $p$, bounds obtained for $r(k, k)$ by these techniques are not too bad (see also Mathon [Ma87]). It is known that $G_p$ shares many properties with a random graph of the same size (e.g., see [GS71], [BT81]). However, it can be shown that infinitely often $G_p$ will contain a complete subgraph of size $c \log p \log\log p$.

### 2.3 Upper bounds

The earliest upper bounds on $r(k, \ell)$, due to Erdös and Szekeres [ES35], follows from the (immediate) recurrence

(2.13) $$r(k, \ell) \leqslant r(k-1, \ell) + r(k, \ell-1)\,.$$

This leads at once to the upper bound

(2.14) $$r(k, \ell) \leqslant \binom{k+\ell-2}{k-1}$$

and, in the special case that $k = \ell$,

(2.15) $$r(k, k) \leqslant \binom{2k-2}{k-1} \sim c \cdot 4^k/\sqrt{k}\,.$$

For fixed values of $\ell$, these bounds have been improved by Graver and Yackel [GY68] to

$$r(k, \ell) \leqslant c\,\frac{k^{\ell-1}\log\log k}{\log k}\,,$$

although a widely quoted upper bound for large values of $k$ and $\ell$, published by Yackel [Y72] now appears to be flawed (cf. [LSW]). In this section we present several recent improved estimates on $r(k, \ell)$. The bound on $r(k, 3)$ (Theorems 2.9, 2.10) and its striking proof is due to Shearer [Sh83], and is based on an earlier argument of Ajtai, Komlós and Szemerédi [AKS80]. We point out that the ideas introduced in [AKS80] have also been applied very effectively by these authors to several other problems as well (cf. [AKS81a], [AKS81b]). The general bound on $r(k, \ell)$ in Theorem 2.11 is due to the second author and represents the first significant improvement on (2.14) in the 50 years since it appeared. In particular, it implies that

$$r(k, \ell) = o\left(\binom{k+\ell-2}{k-1}\right)$$

as $k + \ell \longrightarrow \infty$. Note that (2.13) does not imply $r(k, k) = o\left(\binom{2k}{k}\right)$, even assuming the (unrealistic) boundary conditions $r(j, 3) = 1$ for all $j$.

*Theorem 2. 9* [Sh83]

Let $G$ be a triangle-free graph on $n$ vertices with average degree $d$. Let $\alpha = \alpha(G)$ be the independence number of $G$ and define

$$f(d) := (d\log d - d + 1)/(d-1)^2\,,\ f(0) = 1\,,\ f(1) = 1/2\,.$$

Then

(2.16) $$\alpha \geqslant nf(d)\,.$$

*Proof*: First note that $f$ is continuous for $0 \leqslant d < \infty$ and that $0 < f(d) < 1$, $f'(d) < 0$, $f''(d) > 0$. Further, $f$ satisfies

(2.17) $$(d+1)f(d) = 1 + (d - d^2)f'(d)\,.$$

We will prove (2.16) by induction on $n$. For $n \leqslant d/f(d)$, the theorem clearly holds since the neighbors of any vertex of $G$ must form an independent set, and consequently

$$\alpha \geqslant d \geqslant nf(d)\,.$$

For the vertex $v$ in $G$, let $d_1 = d_1(v)$ denote the degree of $v$. Also, let $d_2 = d_2(v)$ denote the average degree of the neighbors of $v$. We claim that we can always find a vertex $v$ so that

(2.18) $$(d_1+1)f(d) \leqslant 1 + (dd_1 + d - 2d_1d_2)f'(d)$$

holds. To see this, note that the average value of $d_1d_2$ is the same as the average value of $d_1^2$, and is at least $d^2$. Thus, the average value of the RHS of (2.18) is at least as large as the average value of the RHS of (2.17). Since, the average value of the LHS of (2.18) is equal to the LHS of (2.17), then (2.18) holds on the average and consequently, the desired vertex $v$ must exist.

Now, let $G^*$ be formed from $G$ by deleting $v$ and all its neighbors. Of course, $G^*$ is also triangle-free, and has $n - d_1 - 1$ vertices and $\frac{1}{2}nd - d_1 d_2$ edges. Let $d^* = (nd - 2d_1 d_2)/(n - d_1 - 1)$ be the average degree in $G^*$. By the induction hypothesis, $G^*$ contains an independent set of size $(n - d_1 - 1)f(d^*)$. By adding $v$ back to this set, we obtain an independent set in $G$ of size

$$1 + (n - d_1 - 1)f(d^*) .$$

Since $f''(d) \geqslant 0$ for $0 < d < \propto$ we have

$$f'(d) \geqslant f(d) + (d^* - d)f'(d) .$$

Therefore,

$$\begin{aligned}
1 + (n - d_1 - 1)f(d^*) &\geqslant 1 + (n - d_1 - 1)f(d) + (n - d_1 - 1)(d^* - d)f'(d) \\
&\geqslant 1 + (n - d_1 - 1)f(d) + (dd_1 + d - 2d_1 d_2)f'(d) \\
&\geqslant (n - d_1 - 1)f(d) + (d_1 + 1)f(d) \qquad \text{by (2.18)} \\
&= nf(d) .
\end{aligned}$$

Thus, $\alpha \geqslant nf(d)$ as required, and the theorem is proved. ■

As an immediate consequence, we have:

*Theorem 2.10*

$$r(k, 3) < k^2/\log(k/e) . \tag{2.19}$$

*Proof*: Let $G$ be a triangle-free graph on $n$ vertices and suppose $\alpha(G) \leqslant k$. Since the neighbors of any vertex $v$ form an independent set then we must have degree $(v) \leqslant k$. Thus, the *average* degree $d$ in $G$ is at most $k$. Therefore, by (2.16)

$$k \geqslant \alpha(G) \geqslant nf(d) \geqslant nf(k) . \tag{2.20}$$

Hence, if $n > k/f(k)$ then $G$ must contain an independent set of size $k + 1$. This implies

$$R(k + 1, 3) \leqslant 1 + k/f(k) \tag{2.21}$$

which in turn implies (2.19). ■

Theorem 2.9 is a slight extension of an earlier result of Ajtai, Komlós and Szemerédi [AKS80] who proved

$$\alpha(G) \geqslant n/(100d \log d)$$

for any triangle-free graph $G$ with $n$ vertices and average degree $d$. If 3 is replaced by an arbitrary but fixed value $\ell$, then the best bound on $r(k, \ell)$ is given by the following result of Ajtai, Komlós and Szemerédi:

*Theorem 2.11* [AKS80]

$$r(k, \ell) \leqslant (5000)^\ell k^{\ell-1}/(\log k)^{\ell-2} \tag{2.22}$$

for $k$ sufficiently large (depending on $\ell$).

It will be convenient for the next result to define the related quantity $r^*(k, \ell)$, the largest value of $n$ such that there is a red-blue coloring of the edges of $K_n$ having no red $K_k$ and no blue $K_\ell$. Thus,

$$r^*(k, \ell) = r(k+1, \ell+1) - 1 .$$

For arbitrary $k$ and $\ell$ (not satisfying $k >> \ell$ required by Theorem 2.11), the best current upper bound is given by a recent result of the second author:

*Theorem 2.12* [R]

$$r^*(k, \ell) \leqslant c_1 \binom{k+\ell}{k} \Big/ (\log(k+\ell))^{c_2} \tag{2.23}$$

for suitable positive constants $c_i$. The proof of Theorem 2.12 is based on the following result, which is of independent interest.

*Theorem 2.13*

If $\ell = \log k$ then

$$r^*(k, \ell) \leqslant \binom{k+\ell}{k} \Big/ (k+\ell)^c \tag{2.24}$$

for a suitable constant $c > 0$.

Because of space limitations, we will not give the proof of (2.23). Rather, we will illustrate the method used in proving it by proving the following weaker result.

*Theorem 2.14*

$$r^*(k, \ell) < 6 \binom{k+\ell}{k} \Big/ \log\log(k+\ell) \tag{2.25}$$

for $k + \ell$ sufficiently large.

*Proof*: The proof of (2.25) will require the use of several auxiliary lemmas. We first establish these.

*Lemma 2.15* [Go59] Let $G$ be a graph with $n$ vertices and $\beta \binom{n}{2}$ edges, and let $T(X)$ denote the number of triangles in a graph $X$. Then

$$T(G) + T(\overline{G}) \geqslant (\beta^3 + (1-\beta)^3) \binom{n}{3} - \beta(1-\beta) \binom{n}{2} . \tag{2.26}$$

*Proof*: Let $d_1, d_2, \ldots, d_n$ be the degrees of the vertices of $G$. Then

$$\begin{aligned} T(G) + T(\overline{G}) &= \binom{n}{3} - \frac{1}{2} \sum_i (n - d_i - 1) d_i \\ &= \binom{n}{3} - \frac{1}{2} \sum_i ((n-1)d_i - d_i^2) \\ &\geqslant \frac{1}{6} (n(n-1)(n-2) - 3\beta n(n-1)^2 + 3\beta^2 n(n-1)^2) \\ &= (\beta^3 + (1-\beta)^3 \binom{n}{3} - \beta(1-\beta) \binom{n}{2} . \quad \blacksquare \end{aligned}$$

*Lemma 2.16* Suppose $m$ and $n$ satisfy $m > e^2 n > 0$.

Then

$$r^*(m, n) \leq 5 \binom{m+n}{n} \Big/ \log(m/e^2 n) . \tag{2.27}$$

*Proof*: By applying (2.13) repeatedly, we obtain

$$r(m+1, n+1) \leq \sum_{j=0}^{n} \binom{m+n-j-3}{n-3} r(j+1, 3) \tag{2.28}$$

Let $s := m + n$ and $t := \lfloor (s/n)^{1/2} \rfloor - 1$. We break the sum on the RHS of (2.28) into two parts.

$$S_1 := \sum_{j \leq t} \binom{s-j-3}{n-3} r(j+1, 3) \leq \binom{s-3}{n-3} \sum_{j \leq t} \binom{j+2}{2}$$

$$\leq \binom{s}{n} \left(\frac{n}{s}\right)^3 \binom{t+3}{3} . \tag{2.29}$$

Also, by (2.19)

$$S_2 := \sum_{j=t+1}^{s} \binom{s-j-3}{n-3} r(j+1, 3) \leq \sum_{j=t+1}^{s} \binom{s-j-3}{n-3} (j+1)^2 \Big/ \log\left(\frac{j+1}{e}\right)$$

$$\leq \frac{4}{\log(s/e^2 n)} \sum_{j \geq 0} \binom{s-j-3}{n-3} \binom{j+2}{2}$$

$$= \frac{4}{\log(s/e^2 n)} \left\{ \binom{n-1}{2} \binom{s}{n} - (s-2)(n-2) \binom{s-1}{n-1} + \binom{s-2}{2} \binom{s-2}{n-2} \right\}$$

$$= \frac{4}{\log(s/e^2 n)} \left[ 1 - \frac{n(2s-n-1)}{s(s-1)} \right] \binom{s}{n}$$

$$< \frac{4}{\log(s/e^2 n)} \cdot \binom{s}{n} .$$

Thus, using the fact that

$$\left(\frac{n}{s}\right)^3 \binom{t+3}{3} < 1/\log(s/e^2 n) ,$$

we obtain

$$r^*(m, n) < r(m+1, n+1) < S_1 + S_2 < 5 \binom{m+n}{n} \Big/ \log(m/e^2 n)$$

for $m > e^2 n$. ■

*Lemma 2.17* Suppose for some positive $m$ and $n$,

$$r^*(m, n) \geq 16(m+n)^2 . \tag{2.30}$$

Let $s := m + n$ and $\epsilon := \frac{1}{8s} \min(m, n)$. Then at least one of the following holds:

(i) $$r^*(m-1, n) \geqslant \frac{m}{s}\left(1+\frac{\epsilon}{s}\right) r^*(m, n);$$

(ii) $$r^*(m, n-1) \geqslant \frac{n}{s}\left(1+\frac{\epsilon}{s}\right) r^*(m, n);$$

(iii) $$r^*(m-2, n) \geqslant \frac{m(m-1)}{s(s-1)}\left(1+\frac{\epsilon}{s}\right)\left(1+\frac{\epsilon}{s-1}\right) r^*(m, n);$$

(iv) $$r^*(m, n-2) \geqslant \frac{n(n-1)}{s(s-1)}\left(1+\frac{\epsilon}{s}\right)\left(1+\frac{\epsilon}{s-1}\right) r^*(m, n).$$

*Proof:* Set $N := r^*(m, n)$ and let $K_N$ be 2-colored so that no red $K_{m+1}$ and no blue $K_{n+1}$ is formed. Let $d_i^{(r)}$ denote the number of red edges incident to the $i$th vertex of $K_N$, with $d_i^{(b)}$ denoting the analogous quantity for the blue edges. If $d_i^{(r)} \geqslant \frac{m}{s}\left(1+\frac{\epsilon}{s}\right) N$ for some $i$, then by considering the "red" neighbors of this vertex we get (i). Similarly, if $d_i^{(b)} \geqslant \frac{n}{s}\left(1+\frac{\epsilon}{s}\right) N$ then (ii) must hold. Hence, we can assume that

(2.31) $$d_i^{(r)} < \frac{m}{s}\left(1+\frac{\epsilon}{s}\right) N, \quad d_i^{(b)} < \frac{n}{s}\left(1+\frac{\epsilon}{s}\right) N$$

for $1 \leqslant i \leqslant N$. Since

$$d_i^{(r)} + d_i^{(b)} = N - 1$$

then

$$d_i^{(r)} > N - 1 - \frac{n}{s}\left(1+\frac{\epsilon}{s}\right) N$$

(2.32) $$= N\left(1 - \frac{n}{s} - \frac{\epsilon n}{s^2}\right) - 1$$

$$= N\left(\frac{m}{s} - \frac{\epsilon n}{s^2}\right) - 1 .$$

On the other hand, if $\beta$ is defined by

$$\frac{1}{2} \sum_i d_i^{(r)} = \beta \binom{N}{2},$$

then Lemma 2.15 implies that there are either at least

$$A := \beta^3 \binom{N}{3} - \frac{1}{2} \beta(1-\beta) \binom{N}{2}$$

red triangles, or at least

$$B := (1-\beta)^3 \binom{N}{3} - \frac{1}{2} \beta(1-\beta) \binom{N}{2}$$

blue triangles in $K_N$. Consider the first possibility. Then some edge, say $e_0$, is contained in at least

(2.33) $$\frac{3A}{\beta\binom{N}{2}} = \beta^2 N - \frac{3}{2}(1 - \beta) \geqslant \beta^2 N - 2$$

red triangles. Thus, (iii) would follow (by considering the set of vertices of these red triangles, excluding the endpoints of $e_0$) if we could show

(2.34) $$\beta^2 N - 2 \geqslant \frac{m(m-1)}{s(s-1)}\left(1 + \frac{\varepsilon}{s}\right)\left(1 + \frac{\varepsilon}{s-1}\right) N$$

This we now indicate how to do. First, observe that by (2.32)

$$\sum_i d_i^{(r)} = 2\beta\binom{N}{2} > N^2\left(\frac{m}{s} - \frac{\varepsilon n}{s^2}\right) - N$$

so that

(2.35) $$\beta > \frac{\left(N\frac{m}{s} - \frac{\varepsilon n}{s^2}\right) - 1}{N - 1}$$

Thus it will be enough to show that

(2.36) $$\frac{m}{s} - \frac{\varepsilon n}{s^2} + \left(\frac{m}{s} - \frac{\varepsilon n}{s^2} - 1\right)/(N-1) \geqslant$$
$$\frac{m(m-1)}{s(s-1)}\left(1 + \frac{\varepsilon}{s}\right)\left(1 + \frac{\varepsilon}{s-1}\right)^2 N + 2$$

for $N \geqslant 16s^2$

We will not carry out all the details of this computation, which are relatively straight forward (but unenlightening). The basic point is that the important term of (2.36) is

(2.37) $$\left(\frac{m}{s} - \frac{\varepsilon n}{s^2}\right)^2 - \frac{m(m-1)}{s(s-1)}\left(1 + \frac{\varepsilon}{s}\right)\left(1 + \frac{\varepsilon}{s-1}\right)$$

which can be rewritten as

(2.38) $$\left(\frac{m^2}{s^2} - \frac{m(m-1)}{s(s-1)} - \varepsilon\left(\frac{2mn}{s^3} + \left(\frac{1}{s} + \frac{1}{s-1}\right)\frac{m(m-1)}{s(3-1)}\right)\right.$$
$$\left. + \varepsilon^2\left(\frac{n^2}{s^4} - \frac{m(m-1)}{s^2(s-1)^2}\right)\right) N$$

In turn the main contribution to (2.38) turns out to be

(2.39) $$\frac{m^2}{s^2} - \frac{m(m-1)}{s(s-1)} = \frac{mn}{s^2(s-1)}$$

by the definition of s.

The value assigned to $\varepsilon$ now guarantees (2.36), and therefore (2.34), holds.

The case that there are B blue trinagles follows in exactly the same way (using the symmetry of $\beta$ and $1 - \beta$ in the expressions involved) to yield (iv). This completes the proof of Lemma 2.17.

We need one final observation before proceeding to the proof of (2.25).

Fact.

(2.40)
$$\sum_{i=r+1}^{s} 1/i > \log \frac{s+i}{r+1}$$

Proof: The sequence

$$a_n = \sum_{i=i}^{n} 1/i - \log(n+1)$$

is monotone increasing. Consequently

$$\begin{aligned}\sum_{i=r+1}^{s} 1/i &= \sum_{i=i}^{s} 1/i - \sum_{i=i}^{r} 1/i \\ &= \log(s+1) + a_s - \log(r+1) - a_r \\ &> \log \frac{s+1}{r+1}\end{aligned}$$

Proof of Theorem 2.14. Let G be a graph with $r^*(k,\ell)$ vertices. We will consider the following algorithm.

1. Set $m = k$, $n = \ell$, $t = k + \ell$ and

$$x = \frac{1}{8} \frac{\log t}{\log \log t} - 1 .$$

2. If $\max\{m/n, n/m\} \geqslant x$ then *halt*; otherwise go to 3.
3. If $r^*(m, n) < 16(m + n)^2$ then *halt*; otherwise go to 4.
4. Select a pair $(m^*, n^*)$ for which one of the possibilities of Lemma 2.17 occurs. (Thus, $m^* + n^* = m + n - 1$ or $m + n - 2$.) Let $G^*$ be a graph with $r^*(m^*, n^*)$ vertices. Set $m = m^*$, $n = n^*$, $G = G^*$ and go to 2.

Suppose now that the algorithm halts at some graph $G'$ of size $r^*(k', \ell')$. Let $t' := k' + \ell'$ and $p = t - t'$. Note that if $y := \dfrac{\log \log t}{\log t}$ then at each pass through step 4 of the algorithm, the value of $\epsilon = \dfrac{\min(m, n)}{8(m + n)}$ satisfies

$$\epsilon \geqslant \frac{n}{8(m+n)} \geqslant \frac{1}{8(1 + m/n)} \geqslant \frac{1}{8(1+x)} = y \,. \tag{2.41}$$

Hence, by the time we have reached $G'$, we have (by Lemma 2.17) accumulated the "gain factors" to obtain the estimate

$$r^*(k', \ell') = \frac{(k'+1)(k'+2)\cdots(k)(\ell'+1)(\ell'+2)\ldots(scrL)}{(t'+1)(t'+2)\cdots(t)}\left(1 + \frac{y}{t'+1}\right)\cdots\left(1 + \frac{y}{t}\right) r^*(k, \ell)$$

$$= \frac{\binom{t'}{k'}}{\binom{t}{k}} \prod_{i=1}^{p}\left(1 + \frac{y}{t'+i}\right) r^*(k, \ell) \,. \tag{2.42}$$

We now consider several cases, depending on how soon the algorithm halts, and why.

*Case 1.* $t' > \sqrt{t} + 2$.

*Subcase (a).* $r^*(k', \ell') < 16\, t'^2$.

Since $t$ is large then so is $t'$. Thus, by (2.7) we must have $\min(k', \ell') = 2$, say (by symmetry) $\ell' = 2$. Thus,

$$\begin{aligned} r^*(k', \ell') = r^*(k', 2) &\leqslant (k'+1)^2/\log(k'+1)/e) \\ &< (k'+1)^2/\log(\sqrt{t}/e) \\ &= 2(k'+1)^2/\log(t/e^2) \,. \end{aligned} \tag{2.43}$$

Therefore, by (2.42)

$$\begin{aligned} r^*(k, \ell) &\leqslant \frac{\binom{t}{k}}{\binom{k'+2}{2}} r^*(k', \ell') \\ &\leqslant \frac{\binom{t}{k}}{\binom{k'+2}{2}} \cdot \frac{2(k+1)^2}{\log(t/e^2)} < 5\binom{k+\ell}{k}\Big/\log(k+\ell) \end{aligned} \tag{2.44}$$

for $t = k + \ell$ large.

*Subcase (b).* $k'/\ell' \geqslant x = \frac{1}{8} \frac{\log t}{\log\log t} - 1.$

Thus, by Lemma 2.16,

$$r^*(k', \ell') \leqslant 5\binom{k'+\ell'}{k'} \Big/ \log(k'/e^2\ell')$$

so that

$$r^*(k, \ell) \leqslant \frac{5\binom{k+\ell}{k}}{\binom{k'+\ell'}{k'}} \cdot \frac{\binom{k'+\ell'}{k'}}{\log(k'/e^2\ell')}$$

$$\leqslant 5\binom{k+\ell}{k} \Big/ \log\left(\frac{1}{e^2}\frac{\log t}{\log\log t} - 1\right) < 6\binom{k+\ell}{k} \Big/ \log\log(k+\ell) \tag{2.45}$$

for $k + \ell$ sufficiently large.

*Case 2.* $t' \leqslant \sqrt{t} + 2$

Thus,

$$p = t - t' \geqslant t - \sqrt{t} - 2$$

so that the factor in (2.42) we gain is

$$\prod_{i=1}^{p}\left(1 + \frac{y}{t'+i}\right) \geqslant \prod_{i=0}^{t-\sqrt{t}-3}\left(1 + \frac{y}{t-i}\right).$$

$$> y \sum_{i=0}^{t-\sqrt{t}-3} \frac{1}{t-i} + 1$$

$$> y \log \frac{t+1}{\sqrt{t}} \qquad \text{by (2.40)}$$

$$> \frac{1}{2} y \log t = \frac{1}{2} \log\log t .$$

Hence,

$$r^*(k, \ell) \leqslant \frac{\binom{k+\ell}{k}}{\binom{k'+\ell'}{k'}} \cdot \frac{1}{\frac{1}{2}\log\log t} \cdot r^*(k', \ell') < 2\binom{k+\ell}{k} \Big/ \log\log(k+\ell) \tag{2.46}$$

for $k + \ell$ large.

Thus in all cases (2.25) holds and the theorem is proved. ■

## 2.4 Many colors

Up this point, we have investigated the problem of estimating the size of the largest monochromatic clique formed whenever the edges of $K_n$ are 2-colored. In this section we will discuss the case in which $t$ colors are permitted.

Define $r(k_1, k_2, \ldots, k_t)$ to be the least integer $n$ with the property that for every $t$-coloring of the edges of $K_n$, there exists an $i$, $1 \leqslant i \leqslant t$, and a complete subgraph $K_{k_i}$ of $K_n$ having all edges colored by the $i$th color. As before, Ramsey's Theorem guarantees the existence of $r(k_1, \ldots, k_t)$.

The same argument as that used for $r(k, \ell)$ by Erdös and Szekeres gives a general (recursive) upper bound:

*Proposition 2.18*

$$r(k_1, k_2, \ldots, k_t) \leqslant r(k_1 - 1, k_2, \ldots, k_t) + r(k_1, k_2 - 1, \ldots, k_t) + \cdots$$
$$\ldots + r(k_1, k_2, \ldots, k_t - 1) - (t - 2) .$$

*Proof*: Set $n$ equal to the RHS of (2.47) and let the edges of $K_n$ be $t$-colored. Let $C_i$ denote the set of all vertices joined to a given vertex $v$ by an edge having the $i$th color. Then

$$\sum_{i=1}^{t} |C_i| = n - 1$$

and therefore there exists $j$, $1 \leqslant j \leqslant t$, with

$$|C_j| \geqslant r(k_1, \ldots, k_j - 1, \ldots, k_t) .$$

Suppose now that for no $i$, $1 \leqslant i \leqslant t$, does there exist a subset of $C_j$ of size $k_i$ which spans edges having only the $i$th color. Then by the definition of $r(k_1, \ldots, k_j - 1, \ldots, k_t)$, $C_j$ must contain a $K_{k_j-1}$ with all its edges colored by the $j$th color. Adding the vertex $v$ to this $K_{k_j-1}$, we obtain the desired copy of $K_{k_j}$ having all edges with the $j$th color. ■

It follows now by induction on $k_1, \ldots, k_t$ and $t$ that, for example,

$$r(k_1, \ldots, k_t) \leqslant \frac{(k_1 + \cdots + k_t - t)!}{(k_1 - 1)! \cdots (k_t - 1)!} . \qquad (2.48)$$

For $t > 2$, (2.48) can easily be improved by a factor which tends to 0 as $t \longrightarrow \infty$. For a discussion of general lower bounds for $r(k_1, \ldots, k_t)$, the reader is referred to [A72]. For the remainder of the section, we will restrict ourselves to the interesting special case that

$k_1 = \ldots = k_t = 3$. Denote $r(\overbrace{3, 3, \ldots, 3}^{t})$ by $r(3; t)$.

*Theorem 2.19*

For a suitable constant $c > 0$,

$$c(315)^{t/5} \leqslant r(3; t) \leqslant t!(e - 1/24) , t \geqslant 4 .$$

The lower bound is due to Frederickson [Fr79] (cf. [CG83]). For a statement of the upper bound, the reader is referred to the survey paper of Chung and Grinstead [CG83]. Here we will prove only the slightly weaker:

*Theorem 2.20*

$$\frac{3^t + 3}{2} \leqslant r(3; t) \leqslant \lfloor t!\, e \rfloor , t \geqslant 4 . \qquad (2.49)$$

*Proof*: To prove the upper bound, we first show

$$r(3; t) \leqslant t(r(3; t - 1) - 1) + 2 \qquad (2.50)$$

Fix a vertex $v$ of an $r$-colored complete graph $K_n$ with $n = t(r(3; t - 1) - 1) + 2$ vertices. Let $C_i$ denote the set of all vertices $x$ with the property that $\{x, v\}$ is colored by the $i$th color. By the pigeon-hole principle, there exists $j$ with

$$|C_j| \geqslant r(3; t-1) . \tag{2.51}$$

We can assume that no pair in $C_j$ is colored by the $j$th color (since otherwise $x$ together with such a pair forms a triangle in the $j$th color). However, it follows from (2.51) and the definition of $r(3; t-1)$ that $C_j$ contains a monochromatic triangle. Iterating (2.50) now yields the upper bound on (2.49).

To prove the lower bound, let us call a set $A$ of integers *sum-free* if $a, b \in A$ implies $a + b \notin A$. Let $s_t$ denote the largest integer such that the set $\{1, 2, \ldots, s_t\}$ can be partitioned into $t$ sets $A_1, \ldots, A_t$, each of which is sum-free. (A theorem of Schur [Sc16] guarantees that $s_t$ is finite; see Theorem 5.1.)

*Claim.* For any $t > 0$,

$$s_{t+1} \geqslant 3s_t + 1 . \tag{2.52}$$

*Proof:* Let

$$\{1, 2, \ldots, s_t\} = A_1 \cup \cdots \cup A_t$$

be a partition of $\{1, 2, \ldots, s_t\}$ into $t$ sum-free sets. Then the sets

$$B_i = \{3a : a \in A_i\} \cup \{3a - 1 : a \in A_i\} , \quad 1 \leqslant i \leqslant t ,$$

$$B_{t+1} = \{3\ell + 1 : 0 \leqslant \ell \leqslant s_t\} ,$$

are sum-free and moreover, form a partition of $\{1, 2, \ldots, 3s_t + 1\}$. ■

Since $s_1 = 1$, it follows by iterating (2.52) that

$$s_t \geqslant \frac{1}{2}(3^t - 1) . \tag{2.53}$$

Now, set $m = s_t + 1$ and let $C_1 \cup \cdots \cup C_t$ be the $t$-coloring of the edges of $K_m$ with vertex set $\{1, 2, \ldots, m\}$ defined by

$$\{u, v\} \in C_k \text{ iff } |u - v| \in A_k .$$

Suppose this graph contains a monochromatic triangle $\{u, v, w\}$, with $u < v < w$. This implies $v - u, w - v, w - u \in A_k$ for some $k$, contradicting the fact that $A_k$ is sum-free. Thus,

$$m = s_t + 1 \leqslant r(3; t) - 1$$

i.e.,

$$r(3; t) \geqslant \frac{3^t - 1}{2} + 2 = \frac{3^t + 3}{2}$$

as required. ■

The outstanding open problem concerning $r(3; t)$ is to decide whether or not $\lim_{t \to \infty} r(3; t)^{1/t}$ is finite. (The limit is known to exist [Ch73].) Erdős currently is offering $100 for the answer (with a proof!)

## 3. Ramsey numbers for other graphs

During the past 10 years, an impressive number of papers have appeared which investigate analogues of Ramsey's Theorem for other graphs besides complete graphs. An initial motivation for these studies was the hope that progress here might lead to a deeper understanding of the classical (complete graph) problems. While this hope has not yet materialized (for example, the Ramsey numbers are known for all graphs on 5 vertices *except* $K_5$), the subject has developed into a lively and interesting area in its own right. In this section we will describe several of the results (including some very recent ones) which we find particularly attractive.

We begin with a definition. For graphs $G$ and $H$, we let $r(G, H)$ denote the least integer $n$ so that in any coloring of the edges of $K_n$ by red and blue (say), there must always be formed either a red copy of $G$ or a blue copy of $H$. In particular, if $G = K_k$ and $H = K_\ell$ then $r(K_k, K_\ell) = r(k, \ell)$ from the preceding section. When $G = H$, we abbreviate $r(G, G)$ by $r(G)$.

One of the simplest and most general results in this topic is the following theorem of Chvátal and Harary [CH72].

*Theorem 3.1*

(3.1) $$r(G, H) \geqslant (\chi(G) - 1)(c(H) - ) + 1$$

where $\chi(G)$ denotes the chromatic number of $G$ and $c(H)$ denotes the size of the largest connected component of $H$.

*Proof:* Let $m = (\chi(G) - 1)(c(H) - 1)$ and think of $K_m$ as being $\chi(G) - 1$ copies of $K_{c(H)-1}$ with edges interconnecting all pairs of vertices in different copies of $K_{c(H)-1}$. Color all edges within each copy of $K_{c(H)-1}$ blue, and all remaining edges red. Certainly, there is no red copy of $G$ since this would imply that the chromatic number of $G$ is at most $\chi(G) - 1$. On the other hand, there can be no blue copy of $H$ since the largest blue component of $K_n$ has size $c(H) - 1$. Therefore, $r(G, H) > m$. ■

*Corollary 3.2* [C77] For any tree $T_m$ with $m$ vertices,

(3.2) $$r(T_m, K_n) = (m - 1)(n - 1) + 1 .$$

*Proof*: The lower bound follows from (3.1). For $m = 2$ or $n = 2$, (3.2) is immediate. Assume that (3.2) holds for all values of $m'$ and $n'$ with $m' + n' < m + n$. Let $s = (m - 1)(n - 1) + 1$ and consider a 2-colored $K_s$. Let $T'$ be a tree formed from $T$ by the removal of some endpoint $x$ (where $x$ is connected to, say $y$, in $T_m$). By the induction hypothesis, the $K_s$ contains either a *blue* $K_n$ and we are done, or a *red* $T'$. Hence, we may assume the latter holds. Now, remove the $m - 1$ vertices of this red $T'$, leaving a 2-colored $K_{s-(m-1)} = K_{(m-1)(n-2)+1}$. Again, by the induction hypothesis, this graph contains either a red $T_m$ or a blue $K_{n-1}$; we may assume the latter. Thus, in the original $K_s$ we have a red $T'$ and a blue $T_{n-1}$ disjoint from it. Finally, we examine the edges emanating from $y$ to the blue $K_{n-1}$. If any of these edges is red, then we have found a red $T_m$. If none of the edges is red, then $K_{n-1} \cup \{y\}$ forms a blue $K_{n-1}$. ■

Just as in the case of $r(k, \ell)$, it is possible to consider using more than two colors. We let $r(G; t)$ denote the least integer $n$ such that if the edges of $K_n$ are arbitrarily $t$-colored, then a monochromatic copy of $G$ must always be formed. One of the sharper bounds for a number of this type is given by the following result of Chung [Ch74] and Irving [I74] (see also [CG75]). Let $C_4$ denote the cycle on four vertices.

*Theorem 3.3* [I74].

For all $t$,

(3.3) $$r(C_4; t) \leqslant t^2 + t + 1$$

If $t - 1$ is a prime power, then

(3.4) $$r(C_4; t) \geqslant t^2 - t + 2 .$$

Note the dramatic difference between the size of $r(C_4; t)$ and that of $r(C_3; t) = r(3; t)$ given in (2.49) (which grows exponentially in $t$), giving one more example of the difference between the behavior of odd and even cycles in graphs.

Returning again to the case of two colors, one might ask how *slowly* $r(G)$ can grow, as a function of the number of vertices $v(G)$ of $G$. This is answered by the following result of Burr and Erdös.

*Theorem 3.4* [BE76]

If $G$ is a connected graph with $n$ vertices then

$$r(G) \geqslant \lfloor (4n-1)/3 \rfloor . \tag{3.8}$$

Furthermore, for each $n \geqslant 3$, there exist graphs for which equality in (3.8) is achieved.

If $G$ is allowed to be disconnected (but, as always, having no isolated vertices) then $r(G)$ can be much smaller.

*Theorem 3.5* [BE76]

There exist positive constants $c$, $c'$ such that

$$n + \frac{\log n}{\log 2} - c \log \log n \leqslant \min\{r(G) : v(G) = n\} \leqslant n + c'\sqrt{n} . \tag{3.9}$$

It is conjectured that the lower bound in (3.9) is closer to the correct answer.

If $G$ is restricted in various ways, then the Ramsey number can also be restricted. A beautiful example of this behavior is given by a recent result of Chvátal et al. (which was conjectured by Burr and Erdös [BE76]).

*Theorem 3.6* [CRST83]

For each positive integer $d$ there exists a constant $c(d)$ such that for any graph $G$ with $n$ vertices and maximum degree $d$,

$$r(G) \leqslant c(d)n . \tag{3.10}$$

Space limitations prevent us from giving the proof, which uses the powerful regularity lemma of Szemerédi (cf. [Sz76] and Lemma 6.8).

A related result of Beck **[Bec83b]** asserts the following, where we now assume that not only is the maximum degree of $G$ bounded by $d$, but also the chromatic number of $G$ is bounded by $\chi$.

*Theorem 3.7*

If $G$ has $n$ vertices then

$$r(G) < (2n)^{2d^{2\chi-3}} .$$

Burr and Erdös have conjectured a stronger form of Theorem 3.6, which is still unresolved.

*Conjecture 3.8*. For each $d$ there exists a constant $c'(d)$ so that if $G$ is any graph with $n$ vertices and, for any subgraph $G'$ of $G$, the average degree of a vertex in $G'$ is at most $d$, then

$$r(G) \leqslant c'(d)n .$$

An interesting variation of $r(G)$ has been studied by a number of authors recently. This is the *size* Ramsey number, denoted by $r_e(G)$, and defined to be the least integer $m$ for which there exists a graph $H$ with $m$ edges, so that in any 2-coloring of the edges of $H$, a monochromatic copy of $G$ must always be formed.

A beautiful and unexpected result of Beck (settling a $100 problem of Erdös) shows that $r_e$ can be quite small.

*Theorem 3.9* [Bec83a]

If $n$ is sufficiently large then for $P_n$, the path on $n$ vertices,

$$r_e(P_n) < 900n .$$

Beck actually proves a stronger density theorem which implies Theorem 3.9 at once.

*Theorem 3.10* [Bec83a]

If $n$ is sufficiently large, there exists a graph $G$ with fewer than 900 $n$ edges so that any subgraph $H$ of $G$ containing at least half the edges of $G$ must contain a copy of $P_n$.

Beck also considered the corresponding problem for trees $T$ with maximum degree $d$. For this case, he proves:

*Theorem 3.11* [Bec83a]

If a tree $T_n$ has $n$ vertices and maximum degree $d$ then for $n$ sufficiently large,

$$r_e(T_n) < dn(\log n)^{12} . \tag{3.11}$$

We will not give the proofs of Beck's results here (which employ, among other things, the use of the probabilistic method and the Lovász local lemma). The reader is referred to the original papers for the details.

A question left open by Beck was whether the logarithmic term in (3.11) could be replaced by a constant. This was very recently resolved in the affirmative by a striking result of Friedman and Pippenger.

*Theorem 3.12* [FP]

Let $0 < \delta < 1$ and let $d$ be fixed. For every $n$ there is a graph $G$ with $e = O(n)$ edges such that, even after the deletion of all but $\delta e$ edges, $G$ continues to contain every tree with $n$ vertices and maximum degree at most $d$.

The ingenious proof of Theorem 3.12 given in [FP] is definitely non-trivial, and uses earlier results of Beck, Alon and Chung [AC], Lubotzky, Phillips and Sarnak [LPS] and a powerful result of their own implying the universality of expanding graphs with respect to small trees. More precisely, this last result is:

*Theorem 3.13* [FP]

If $H$ is a nonempty graph so that every subset of size $x \leqslant 2n - 2$ vertices has at least $(d + 1)x$ neighbors then $H$ contains every tree with $n$ vertices and maximum degree at most $d$.

The restriction here on the maximum degree is necessary, as Beck [Bec83a] observes, by showing that there are trees $T_n$ with $n$ vertices for which $r_e(T_n) \geqslant n^2/8$. By contrast, any tree $T_n'$ with $n$ vertices satisfies $r(T_n) \leqslant 4n + 1$ (see [EG73]).

For general graphs $G$ with $n$ vertices and maximum degree $d$, it has been shown by Rödl and Szemerédi [RS] that

$$r_e(G) = o(n^2) .$$

On the other hand, there exist graphs $G$ with maximum degree 3 having

$$r_e(G) > n(\log n)^c$$

for an absolute constant $c > 0$.

We conclude this section with several rather nice open problems in this area.

*Conjecture 3.14.* (Erdös) For some $\epsilon > 0$,

$$r(C_4, K_n) = o(n^{2-\epsilon}) .$$

*Conjecture 3.15.* (Erdös) If $G$ has $\binom{n}{2}$ edges then

$$r(G) \leqslant r(K_n) .$$

More generally, if $G$ has $\binom{n}{2} + t$ edges, $0 \leqslant t \leqslant n$, then

$$r(G) \leqslant r(K_n(t))$$

where $K_n(t)$ denotes the graph formed by connecting a new vertex to $t$ of the vertices of a $K_n$.

*Conjecture 3.16.* (Erdös-Graham [EG73])

$$r(C_5;t)/r(C_3;t) \longrightarrow 0$$

as $t \longrightarrow \infty$.

Define the *induced* Ramsey number $r^*(G)$ of a graph $G$ to be the least integer $m$ for which there exists a graph $H$ with $m$ vertices so that in any 2-coloring of the edges of $H$, there is always an *induced* monochromatic copy of $G$ in $H$. The existence of $r^*(G)$ was shown independently by Deuber [D75], Erdös, Hajnal and Pósa [EHP75], and Rödl [R73].

*Problem 3.17.* If $G$ has $n$ vertices, is it true that

$$r^*(G) < c^n$$

for some absolute constant $c$?

This can be shown to hold when $G$ is bipartite (using techniques from [R73]). It is also known that for general graphs $G$,

$$r^*(G) < 2^{2^{n^{1+o(1)}}} .$$

*Problem 3.18.* Is it true that

$$r^*(P_n) < cn \ ?$$

*Problem 3.19.* (Trotter) Is it true that for each $d$ there is a $c(d)$ such that if $G_n$ has $n$ vertices and maximum degree $d$ then $r^*(G_n) \leqslant n^{c(d)}$?

*Problem 3.20.* (Erdös) Is it true that if $G$ has $e$ edges then

$$r(G) < 2^{ce^{1/2}} \qquad (3.12)$$

for some absolute constant $c$? If true, then (3.12) would be, apart from the value of the constant, best possible.

## 4. Hypergraphs

In this section we consider extensions of the preceding questions to the general setting of hypergraphs. More precisely, set $[n] = \{1, 2, \ldots, n\}$ and

$$[n]^p = \{A: A \subseteq [n], |A| = p\} .$$

We will consider colorings of $[n]^p$ where $p$ is an arbitrary integer. This extends the case $p = 2$ considered earlier.

*Definition.* The symbol $n \longrightarrow (\ell_1, \ell_2, \ldots, \ell_t)^p$ will denote the validity of the following statement: For every $t$-coloring of $[n]^p$ there exists $i$, $1 \leqslant i \leqslant t$, and a set $T$, $|T| = \ell_i$ so that $[T]^p$ is colored only by color $i$.

If $\ell_1 = \ell_2 = \ldots = \ell_t = \ell$, then this will be abbreviated by writing $n \longrightarrow (\ell)_t^p$. The general Ramsey numbers are defined as follows:

$$r_p(\ell_1, \ell_2, \ldots, \ell_t) = \min\{n_0: n \longrightarrow (\ell_1, \ell_2, \ldots, \ell_t)^p \text{ for } n \geqslant n_0\} ,$$

$$r_p(\ell, t) = \min\{n_0: n \longrightarrow (\ell)_t^p \text{ for } n \geqslant n_0\}$$

$$r_p(\ell) = r_p(\ell, 2) .$$

Recall that in Section 1 we gave a proof that

(4.1) $$(\sqrt{2}+o(1))^\ell \leqslant r_2(\ell) \leqslant (4+o(1))^\ell$$

so that the growth of the function $r_2(\ell)$ is exponential in $\ell$. For $p \geqslant 3$, much less is known about the order of magnitude of the function $r_p(\ell)$. An upper bound follows from the proof of Ramsey's theorem.

*Theorem 4.1* [ER52]

(4.2) $$\mathrm{Log}_{p-1}(r_p(\ell)) \leqslant c_p \ell$$

where $\mathrm{Log}_{p-1}$ denotes the $(p-1)$-fold iterated logarithm.

*Proof*: The proof proceeds by induction on $p$. For $p = 2$, (4.2) follows from (4.1), while for $p = 1$, (4.2) is just the pigeon-hole principle. Suppose (4.2) holds for $p - 1$, so that $r_{p-1}(\ell)$ exists and satisfies (4.2). Set

$$u = r_{p-1}(\ell - 1) + 1$$

and define the numbers $x_{u-1}, \ldots, x_{p-2}$ by

$$x_{u-1} = 1, \text{ and for } p - 2 \leqslant i < u - 1 ,$$

$$x_i = x_{i+1} 2^{\binom{i+1}{p-1}} + 1 .$$

Let $n = x_{p-2} + (p - 2)$. We prove

(4.3) $$r_p(\ell) \leqslant n .$$

Suppose now that $[n]^p$ is 2-colored, say by the coloring $\chi$. Select distinct points $v_1, v_2, \ldots, v_{p-2}$ in $[n]$ and define

$$V_{p-2} = [n] - \{v_1, v_2, \ldots, v_{p-2}\} .$$

In general, now, suppose $v_1, v_2, \ldots, v_i$ and $V_i$ have been defined. We proceed as follows:

($\alpha$) Select $v_{i+1} \in V_i$ arbitrarily;

($\beta$) Partition $V_i - \{v_{i+1}\}$ into equivalence classes by defining

$$v \equiv v' \textit{ iff}$$

for every choice of $v_{j_1}, v_{j_2}, \ldots, v_{j_{p-1}} \in \{V_i, \ldots, V_{i+1}\}$ the sets $\{v_{j_1}, v_{j_2}, \ldots, v_{j_{p-1}}, v\}$ and $\{v_{j_1}, v_{j_2}, \ldots, v_{j_{p-1}}, v'\}$ have the same color;

($\gamma$) Define $V_{i+1}$ to be the set of those $v$ belonging to the largest equivalence class in ($\beta$).

Thus,

$$|V_{i+1}| \geqslant (|V_i| - 1)2^{-\binom{i+1}{p-1}} \geqslant x_{i+1} .$$

We continue until $v_1, v_2, \ldots, v_u$ are constructed. This is possible since $V_i$ is nonempty for $i = p-2, \ldots, u-1$ (and $|V_i| \geqslant x_i \geqslant 1$). The sequence $v_1, v_2, \ldots, v_u$ therefore has the following property: The color of $\{v_{i_1}, v_{i_2}, \ldots, v_{i_{p-1}}, v_{i_p}\}$, $i_1 < i_2 < \ldots < i_p$, is not changed if $v_{i_p}$ is replaced by any $v_j$ where $j > i_{p-1}$. In other words, the color of any (ordered) $p$-set depends only its first $p-1$ elements. Let $\chi^*$ be the 2-coloring of the $(p-1)$-element subsets of $\{v_1, v_2, \ldots, v_{u-1}\}$ defined by

$$\chi^*\{v_{i_1}, v_{i_2}, \ldots, v_{i_{p-1}}\} = \chi\{v_{i_1}, v_{i_2}, \ldots, v_{i_{p-1}}, v_u\} .$$

By the choice of $u = r_{p-1}(\ell - 1) + 1$ we obtain a $\chi^*$-monochromatic $\ell$-set $\{v_{i_1}, v_{i_2}, \ldots, v_{i_{\ell-1}}, v_u\}$ (which, of course, is also $\chi$-monochromatic). This proves (4.3). Since

$$u = r_{p-1}(\ell - 1) + 1 \leqslant r_{p-1}(\ell)$$

then by induction

$$\operatorname{Log}_{p-2}(u) \leqslant \operatorname{Log}_{p-2}(r_{p-1}(\ell)) \leqslant c_{p-1}\ell .$$

On the other hand,

$$n < \prod_{i=p-2}^{u-1} \left(2^{\binom{i+1}{p-1}} + 1\right) < 2^{2^{\binom{u}{p}}} .$$

Therefore,

$$\operatorname{Log}_{p-1}(r_p(\ell)) \leqslant \operatorname{Log}_{p-2}\left[2\binom{u}{p}\right] \leqslant c_p \ell$$

as required. ■

We next discuss lower bounds on $r_p(\ell)$.

*Theorem 4.2*

$$r_3(\ell) \geqslant \frac{\ell}{e} \cdot 2^{\ell^2/6}(1 + o(1)) . \tag{4.4}$$

*Proof*: The proof of this result is quite similar to that of Theorem 2.1, so we will only give a sketch. Consider a random 2-coloring of the set $[n]^3$. The probability that a particular $\ell$-set is monochromatic is $2^{1-\binom{\ell}{3}}$. Thus, if

$$\binom{n}{\ell} 2^{1-\binom{\ell}{3}} < 1$$

then $n < r_3(\ell)$. A simple computation now yields (4.4). ■

Although the next theorem is a major tool in establishing lower bounds for $r_p(\ell)$, $p \geqslant 4$, we will not give a proof. Instead, we will present a similar, but much less technical proof, of a related result (Theorem 4.5).

*Theorem 4.3* [EHR65] (Stepping-Up Lemma)

If $n \not\rightarrow (\ell)_2^p$ for $p \geqslant 3$ then

$$2^n \not\rightarrow (2\ell + p - 4)_2^{p+1}$$

(where $m \not\rightarrow (k)$ indicates that $m \longrightarrow (k)$ does not hold). This together with Theorem 4.2 implies

*Theorem 4.4*

For $p \geqslant 3$,

$$\text{Log}_{p-2}(r_p(\ell)) \geqslant c_p' \ell^2 . \qquad (4.5)$$

The asymptotic behavior of $r_p(\ell)$ is not known for $p \geqslant 3$. However, because of the Stepping-Up Lemma, any improvement on the lower bound of $r_3(\ell)$ would yield a corresponding improvement on the lower bounds for $r_p(\ell)$, $p > 3$. The major open problem here (and indeed, one of the main unsolved problems in Ramsey theory) is the determination of the order of growth of $r_3(\ell)$. P. Erdös is currently offering $500 for an answer to the following problem.

*Problem 4.5.* Is there an absolute constant $c > 0$ such that

$$\log\log r_3(\ell) \geqslant c\ell \, ? \qquad (4.6)$$

It is interesting to note that if *four* colors are allowed (rather than two), then the analogue to (4.6) is valid, i.e.,

$$\log\log r_3(\ell,4) \geqslant c\ell . \qquad (4.7)$$

This is a consequence of the next result.

*Theorem 4.5* (Hajnal; cf. [EHMR84], Th. 26.3)

If $n \not\rightarrow (\ell)_2^2$ then $2^n \not\rightarrow (\ell + 1)_4^3$.

*Proof*: Let $[n]^2 = C_1 \cup C_2$ be a 2-coloring with no monochromatic $\ell$-set, and let $<$ be the lexicographic order on $\mathscr{P}(n)$, the power set of $[n]$, given by:

$$a_1 < a_2 \textit{ iff } \max(a_1 - (a_1 \cap a_2)) < \max(a_2 - (a_1 \cap a_2)) .$$

For $a_1 \neq a_2 \in \mathscr{P}(n)$, we also define

$$\delta(a_1, a_2) = \max\{i\colon\ i \in (a_1 - a_2) \cup (a_2 - a_1)\} .$$

For $a_1, a_2, a_3 \in \mathscr{P}(n)$ with $a_1 < a_2 < a_3$, let

$$\delta_1 = \delta(a_1, a_2),\ \delta_2 = \delta(a_2, a_3) .$$

Finally, set

$$\{a_1,a_2,a_3\} \in S_1 \textit{ iff } \{\delta_1,\delta_2\} \in C_1 \text{ and } \delta_1 < \delta_2,$$

$$\{a_1,a_2,a_3\} \in S_2 \textit{ iff } \{\delta_1,\delta_2\} \in C_1 \text{ and } \delta_1 > \delta_2,$$

$$\{a_1,a_2,a_3\} \in S_3 \textit{ iff } \{\delta_1,\delta_2\} \in C_2 \text{ and } \delta_1 < \delta_2,$$

$$\{a_1,a_2,a_3\} \in S_4 \textit{ iff } \{\delta_1,\delta_2\} \in C_2 \text{ and } \delta_1 > \delta_2.$$

Suppose now that there is a family $X \subseteq \mathscr{P}(n)$, $|X| = \ell + 1$, which is monochromatic. Assume that $[X]^3 \subseteq S_1$ (the other three cases are similar). Write

$$X = \{a_1,a_2,\ldots,a_{\ell+1}\}, \quad a_1 < a_2 < \ldots < a_{\ell+1}$$

and let $\delta_i := \delta(a_i,a_{i+1})$, $1 \leqslant i \leqslant \ell$. For $i \leqslant \ell - 1$, $\{a_i,a_{i+1},a_{i+2}\} \in S_1$ so that

$$\delta_i = \delta(a_i,a_{i+1}) < \delta(a_{i+1},a_{i+2}) = \delta_{i+1}.$$

Thus,

$$\delta_1 < \delta_2 < \ldots < \delta_\ell .$$

However, for arbitrary $1 \leqslant i < j \leqslant \ell$, $\{a_i,a_{i+1},a_{j+1}\} \in S_1$ and consequently,

$$\{\delta(a_i,a_{i+1}),\delta(a_{i+1},a_{j+1})\} = \{\delta_i,\delta_j\} \in C_1$$

(where $\delta(a_{i+1},a_{j+1}) = \delta(a_j,a_{j+1}) = \delta_j$ follows from the monotonicity of the $a_k$'s). This implies that $\{\delta_1,\delta_2,\ldots,\delta_\ell\}$ is monochromatic, contradicting our hypothesis on the coloring of $[n]^2$. ∎

Applying Theorem 4.5 with the estimate of Theorem 2.1, we obtain:

*Corollary 4.6.* For $\ell$ sufficiently large,

$$\log\log r_3(\ell,4) > \frac{1}{2}\ell \log 2 .$$

We conclude this section with several remarks. Theorem 4.1 is a quantitative form of Ramsey's theorem. The upper bounds on $r_p(\ell)$ are due to Erdös and Rado [ER52]. The lower bounds on $r_p(\ell)$ are due to Erdös and Hajnal [EH72] (cf. [GRS80]).

In summary, the order of growth of the function $r_3(\ell)$ is not known. However, if we allow $t \geqslant 4$ colors, then we do know that $r_3(\ell;t)$ is doubly exponential in $\ell$. If just *three* colors are allowed, there is a modest improvement of (4.4) due to Erdös and Hajnal [EH]:

(4.8) $$r_3(\ell;3) > \exp(c\ell^2\log^2\ell) .$$

Finally, we state one more related problem which was considered in [EH72].

For fixed $n,\ell,u,v$ and $p$, the notation

(4.9) $$n \longrightarrow \left(\ell, \begin{bmatrix} u \\ v \end{bmatrix}\right)^p$$

will denote the truth of the following statement: For any red-blue coloring of $[n]^p$, either there is an $\ell$-set $X \subseteq [n]$ with all elements of $[X]^p$ red, or there is a $u$-set $Y \subseteq [n]$ with at least $v$ elements of $[Y]^p$ blue. Let $f_p(n, u, v)$ denote the largest value of $\ell$ for which (4.9) holds. The conjecture of Erdös and Hajnal concerns the behavior of $f_p(n, u, v)$ as $v$ increases from 1 to $\binom{u}{p}$. It asserts that there exist

$$1 < v_1 < v_2 < \ldots < v_{p-2} \leqslant \binom{u}{p}, \quad \text{where } v_i = v_i(u)$$

(but is independent of $n$) such that the function $f_p(n, u, v)$ grows like a power of $n$ for $v \in [1, v_1 - 1]$, like a power of $\log n$ for $v \in [v_1, v_2 - 1]$, like a power of $\log\log n$ for $v \in [v_2, v_3 - 1]$, etc., and finally, like a power of $\mathrm{Log}_{p-2} n$ for $v \in [v_{p-2}, \binom{u}{p}]$.

## 5. The Theorems of Schur and Rado-Folkman-Sanders

Set $m = \lfloor t!e \rfloor$ and suppose that $[1, m] = C_1 \cup C_2 \cup \ldots \cup C_t$ is a $t$-coloring of the integers in the interval $[1, m]$. Consider the induced $t$-coloring of the edges of the complete graph $K_{m+1}$ with vertex set $\{1, 2, \ldots, m+1\}$ where the edge $\{u, v\}$ is colored by the $i$th color *iff* $|u - v| \in C_i$. By Theorem 2.20, we must find a monochromatic triangle $\{u, v, w\}$ with $u < v < w$. However, this means that the integers $x = v - u$, $y = w - v$ and $z = w - u = x + y$ all belong to the same $C_j$ for some $j$.

This argument yields the quantitative form of a theorem which was proved some 70 years ago by I. Schur:

*Theorem 5.1* [Sc16]

Suppose

$$\overline{m} \geqslant r(3; t) - 1 \geqslant \lfloor t!e \rfloor$$

and the set of integers $\{1, 2, \ldots, \overline{m}\}$ is $t$-colored. Then there exist integers $x, y, z \in \{1, 2, \ldots, \overline{m}\}$ having the same color such that $x + y = z$.

Recall (cf. the proof of Theorem 2.20) that the minimum $\overline{m}$ for which Theorem 5.1 holds was denoted by $s_t + 1$. By Theorem 2.19, $s_t \geqslant c(315)^{t/5}$ for a suitable positive $c$. By replacing $(315)^{1/5} = 3.16\ldots$ by a slightly larger value, it is possible to guarantee that $x$ and $y$ are distinct. We shall now deal with this problem in a somewhat more general form:

Let $X = \{x_1, x_2, \ldots, x_k\}$ be a set of integers. Denote by $\Sigma X$ the set

$$\{\sum_{i \in I} x_i\colon\ \varnothing \neq I \subseteq \{1, 2, \ldots, k\}\}$$

of all $2^k - 1$ sums of nonempty subsets of the $x_i$. Of course, these sums do not all have to be distinct but, in general, $|\Sigma X| \leqslant 2^k - 1$ for any $k$-element set $X$.

Our main concern here will be to discuss the following two theorems:

*Theorem 5.2* (Non-Repeated Sums Theorem).

For any choice of positive integers $k$ and $t$ there is a least integer $S(k, t)$ so that if $n \geqslant S(k, t)$ and $\{1, 2, \ldots, n\}$ is $t$-colored, say $\{1, 2, \ldots, n\} = C_1 \cup \ldots \cup C_t$, then there exists a set $X$ with $|X| = k$ so that $|\Sigma X| = 2^k - 1$ and $\Sigma X \subseteq C_i$ for some $i$.

*Theorem 5.3* (Disjoint Unions Theorem).

For any choice of positive integers $k$ and $t$ there is a least integer $\mathrm{U}(k, t)$ so that if $n \geqslant \mathrm{U}(k, t)$ and $\{1, 2, \ldots, n\}$ is $t$-colored, say $\{1, 2, \ldots, n\} = C_1 \cup \ldots \cup C_t$, then there exists a collection of $k$ pairwise disjoint nonempty sets $Z_1, Z_2, \ldots, Z_k \subseteq \{1, 2, \ldots, n\}$ so that all nonempty unions of the form $\bigcup_{i \in I} Z_i$, $\varnothing \neq I \subseteq \{1, 2, \ldots, k\}$, are in a single $C_j$ for some $j$.

The non-repeated sums theorem was proved by Rado [Ra33], Folkman (unpublished; cf. [GR71b]) and Sanders [Sa68]; the disjoint unions theorem was derived by Graham and Rothschild [GR71a] as a consequence of a general Ramsey theorem for $n$-parameter sets.

The following proposition is part of the folklore.

*Proposition 5.4*

$$\log_2 S(k,t) \leqslant \cup(k,t) < \binom{S(k,t)+1}{2}. \tag{5.1}$$

*Proof*: First we prove the upper bound. Partition a set $X$ of cardinality $\binom{S(k,t)+1}{2}$ into blocks $X_j$, $1 \leqslant j \leqslant S(k,t)$, satisfying $|X_j| = j$. Suppose that all the subsets of $X$ are $t$-colored (and therefore, so are the sets $X_j$). Assign to each integer $j \leqslant S(k,t)$ the color that $X_j$ has, and apply Theorem 5.2. This immediately yields the required sets $Z_1, Z_2, \ldots, Z_k$.

Now we prove the lower bound. Suppose that $\{1, 2, \ldots, 2^{\cup(k,t)}\}$ is $t$-colored. This induces a $t$-coloring of the nonempty subsets of $\{1, 2, \ldots, \cup(k,t)\}$ by assigning to the set $J \subseteq \{1, 2, \ldots, \cup(k,t)\}$ the same color as that assigned to the integer $\sum_{j \in J} 2^j$. By Theorem 5.3 we must have $k$ pairwise disjoint sets $Z_1, Z_2, \ldots, Z_k$ with all unions monochromatic. Set $x_i = \sum_{z \in Z_i} 2^z$, $1 \leqslant i \leqslant k$, and $X = \{x_1, x_2, \ldots, x_k\}$. Clearly the set $\Sigma X$ is monochromatic and $|\Sigma X| = 2^k - 1$, which implies $S(k,t) \leqslant 2^{\cup(k,t)}$, as required. ■

A. Taylor [T81] has given upper bounds for $S(k,t)$ and $\cup(k,t)$ which are of the form of a many times iterated exponential. More precisely, for integers $p$ and $q$, define:

$$T_p(1) = p\,,$$

$$T_p(q+1) = p^{T_p(q)}\,.$$

Thus, $T_p(q)$ is a tower of $p$'s of height $q$.

*Theorem 5.5*

$$S(k,2) \leqslant T_3(4k-1)\,, \quad \cup(k,2) \leqslant T_3(4k-2)\,. \tag{5.2}$$

In [T81], Taylor also gives similar upper bounds for the general values $S(k,t)$ and $\cup(k,t)$ in which both the terms and the heights of the towers now depend on $k$ and $t$.

Because of the lower bound in Proposition 5.4, the first inequality of (5.2) is a consequence of the second one. We will give a proof of the second inequality here which is based in part on ideas from [R82] and [Fu85], and is slightly different from the proof of Taylor.

The following result is a special case of a more general theorem of Erdös [E65].

*Lemma 5.6*. Let $A_1, A_2, \ldots, A_r$ be sets with cardinalities $a_1, a_2, \ldots, a_r$, respectively, which satisfy:

$$a_1 = t + 1$$

and

$$a_j = t \prod_{i=1}^{j-1} \binom{a_i}{2} + 1\,, \quad j = 2, \ldots, r\,. \tag{5.3}$$

Suppose that the set $A_1 \times A_2 \times \ldots \times A_r$ is $t$-colored. Then there exist sets $B_i \subseteq A_i$ with $|B_i| = 2$ for $1 \leqslant i \leqslant r$, such that the set $B_1 \times B_2 \times \ldots \times B_r$ is monochromatic.

*Proof*: We will proceed by induction on $r$. For $r = 1$, the statement is immediate. Suppose the statement has been proved for $t = j - 1$ and assume that $A_1 \times A_2 \times \ldots \times A_j$ has been $t$-colored. By induction, for each $a \in A_j$ there exist sets $B_i(a) \subseteq A_i$, $|B_i(a)| = 2$, such that $\prod_{i=1}^{j-1} B_i(a) \times \{a\}$ is monochromatic. Since

$$a_j = t \prod_{i=1}^{j-1} \binom{a_i}{2} + 1$$

must exist distinct $a, a' \in A_j$ having $B_i(a) = B_i(a') := B_i$, and moreover, so that the set

$$\prod_{i=1}^{j-1} B_i \times \{a\} \cup \prod_{i=1}^{j-1} B_i \times \{a'\} = (\prod_{i=1}^{j-1} B_i) \times \{a, a'\}$$

is monochromatic. ■

*Remark*: Note that for $t = 2$, (5.3) gives

$$a_1 = 3\ ,\ a_2 = 7 \text{ and for } j \geqslant 3\ ,$$

$$a_j \leqslant 2 \prod_{i=1}^{j-1} \binom{a_i^2}{2} \leqslant \frac{1}{2} \cdot 3^{\sum_{i=1}^{j-1} 2\cdot 3^i} < \frac{1}{2} \cdot 3^{3^j}\ .$$

Thus,

(5.4) $$\sum_{j=1}^{r} a_j < 3^{3^{r-1}} + \frac{1}{2} \cdot 3^{3^r} < 3^{3^r}\ .$$

*Definition.* Let $\mathscr{X} = \{X_1, X_2, \ldots, X_n\}$ be a collection of pairwise disjoint sets. By $\cup(\mathscr{X})$ we mean the set of all unions of the $X_i$, i.e.,

$$\cup(\mathscr{X}) := \{\bigcup_{i \in I} X_i\colon\ I \subseteq \{1, 2, \ldots, n\}\}\ .$$

*Lemma 5.7.* Let $r$ and $t$ be fixed integers and let $a_1, a_2, \ldots, a_r$ be defined by (5.3). Further, let $\mathscr{X}$ be a collection of $m := \sum_{j=1}^{r} a_j$ disjoint sets $X_1, X_2, \ldots, X_m$, and suppose that the set $\cup(\mathscr{X})$ is $t$-colored. Then there exist pairwise disjoint nonempty sets $Y_0, Y_1, \ldots, Y_r \in \cup(\mathscr{X})$ so that the set

(5.5) $$\{Y_0 \cup Y\colon\ Y \in \cup(\{Y_1, Y_2, \ldots, Y_r\})$$

is monochromatic.

*Proof*: For the given values of $r$ and $t$, let $A_1, A_2, \ldots, A_r$ be the sets, and $a_1, a_2, \ldots, a_r$ be the numbers, from Lemma 5.6. Now, let $\cup(\mathscr{X}) = C_1 \cup C_2 \cup \ldots \cup C_t$ be a $t$-coloring. We will find the desired monochromatic set of the (5.5) as a consequence of Lemma 5.6.

Let $\phi\colon A_1 \cup A_2 \cup \ldots \cup A_r \longrightarrow \mathscr{A}$ be a bijection. Suppose that the elements of each of the sets $A_i$ are linearly ordered by $\leqslant_i$ for $i = 1, 2, \ldots, r$. To each $r$-tuple $(v_1, v_2, \ldots, v_r) \in A_1 \times A_2 \times \ldots \times A_r$ we assign the set $\bigcup_{i=1}^{r} \bigcup_{u_i \leqslant_i v_i} \phi(u_i)$. Consider the $t$-coloring

$$A_1 \times A_2 \times \ldots \times A_r = C_1' \cup C_2' \cup \ldots \cup C_t'$$

defined by

$$(v_1, v_2, \ldots, v_r) \in C_j' \textit{ iff } \bigcup_{j=1}^{r} \bigcup_{u_i \leqslant_i v_i} \phi(u_i) \in C_j\ .$$

Since the sets of the form (5.5) correspond to sets $B_1 \times B_2 \times \ldots \times B_r$, $B_i \subseteq A_i$, $|B_i| = 2$, then Lemma 5.7 follows at once from Lemma 5.6. ■

Note that when $t = 2$ then by (5.4) we have $m < 3^{3^r}$.

*Proof of Theorem 5.5.* We will apply Lemma 5.7 iteratively. Let $m = T_3(4k-2)$ and suppose that the power set $\mathscr{P}(Z)$ of some $m$-element set $Z$ is 2-colored. By Lemma 5.7 we can find pairwise disjoint nonempty sets $Y_0^{(1)}, Y_1^{(1)}, \ldots, Y_{m_1}^{(1)}$ with $m_1 = T_3(4k-4)$ such that the set

$$\{Y_0^{(1)} \cup Y^{(1)}: Y^{(1)} \in \cup(\{Y_1^{(1)}, Y_2^{(1)}, \ldots, Y_{m_1}^{(1)}\})$$

is monochromatic. Now apply Lemma 5.7 again to obtain nonempty pairwise disjoint sets

$$Y_0^{(2)}, Y_1^{(2)}, \ldots, Y_{m_2}^{(2)} \in \cup(\{Y_1^{(1)}, Y_2^{(1)}, \ldots, Y_{m_1}^{(1)}\})$$

with $m_2 = T_3(4k-6)$ so that the set

$$\{Y_0^{(2)} \cup Y^{(2)}: Y^{(2)} \in \cup(\{Y_1^{(2)}, Y_2^{(2)}, \ldots, Y_{m_2}^{(2)}\})\}$$

is monochromatic, etc.

This process is continued until we finally obtain nonempty pairwise disjoint sets

$$Y_0^{(2k-1)}, Y_1^{(2k-1)} \in \cup(\{Y_1^{(2k-2)}, Y_1^{(2k-2)}, \ldots, Y_{m_{2k-2}}^{(2k-2)}\})$$

(since $m_{2k-1} = T_3(0) = 1$) so that the set

$$\{Y_0^{(2k-1)} \cup Y^{(2k-1)}: Y^{(2k-1)} \in \cup(\{Y_1^{(2k-1)}\})$$

is monochromatic.

Now, since each of the sets $Y_0^{(i)}$, $i = 1, 2, \ldots, 2k-1$, has one of two colors, then there must exist indices $i_1 < i_2 < \ldots < i_k$ such that all the sets $Y_0^{(i_j)}$, $1 \leqslant j \leqslant k$, have the same color. However, by the construction of the $Y_0^{(i)}$, the color of any (nonempty) set $Y = Y_0^{(s_1)} \cup Y_0^{(s_2)} \cup \ldots \cup Y_0^{(s_\ell)}$ with $s_1 < s_2 < \ldots < s_\ell$ depends only on the order of $Y_0^{(s_1)}$. Thus, $\cup(\{Y_0^{(i_1)}, Y_0^{(i_2)}, \ldots, Y_0^{(i_k)}\})$ is monochromatic and the theorem is proved. ■

If our state of knowledge in the hypergraph case of Ramsey's theorem is considered unsatisfactory (we do not even know the order of growth of the Ramsey function $r_3(\ell)$), here the situation is much worse. While the upper bounds we have derived are expressed in terms of multiply-iterated exponential functions, the best lower bounds currently known are incomparably smaller.

*Theorem 5.8*

$$S(k, 2) \geqslant \frac{k^2}{e^2} \cdot 2^{\frac{1}{k} \cdot 2^k} \tag{5.6}$$

and

$$\cup(k, 2) \geqslant 2^k/\log 2k \; .$$

The proofs are based on a standard use of the probabilistic method (cf. [ES74]). The lower bound for $\cup(k, 2)$ is stated in [T81]. Here, we give only the proof of the first inequality.

Let us call a set $\{x_1, x_2, \ldots, x_k\} \subseteq [1, n]$ *good* if

$$|\Sigma\{x_1, x_2, \ldots, x_k\}| = 2^k - 1 \; .$$

To each such set associate the set of all possible sequences $(x_{i_1}, x_{i_1} + x_{i_2}, \ldots, x_{i_1} + x_{i_2} + \ldots + x_{i_k})$. It is clear that:

(i) Each sequence is associated with at most one set;

(ii) A good set is associated with $k!$ different sequences.

Thus, since there are at most $\binom{n}{k}$ possible sequences (they are all increasing) then there are at most

$\frac{1}{k!}\binom{n}{k}$ good sets.

Now 2-color the integers in $[1, n]$ randomly so that each integer is independently colored red or blue with probability 1/2. Then, if

$$\frac{1}{k!}\binom{n}{k}2^{-2^k+2} < 1 \tag{5.7}$$

then there must exist some 2-coloring without a monochromatic set of the form $\Sigma X$ satisfying $|X| = k$ and $|\Sigma X| = 2^k - 1$. A simple computation shows that (5.7) holds provided

$$n \leqslant \frac{k^2}{e^2} \cdot 2^{\frac{1}{k} \cdot 2^k}$$

and the assertion is proved. ■

If not all the subset sums are required to be distinct, then the corresponding quantity $S^*(k, 2)$ is only known (by unpublished results of Erdös and Spencer) to satisfy

$$S^*(k, 2) > \exp(ck^2/\log k) .$$

As we noted at the beginning of this section, one can give an upper bound for $S(3, t)$ of the form $c^{t \log t}$. It would be very interesting to know if a similar "small" upper bound exists for $S(4, t)$.

## 6. van der Waerden's Theorem

The celebrated theorem of van der Waerden on arithmetic progressions forms a cornerstone in the edifice of Ramsey theory. In this section we will discuss various numerical aspects of this result. For completeness, we also give a statement and short proof of the theorem.

*Theorem 6.1* (van der Waerden [W27]).

For every pair of integers $k$ and $r$, there exists a least integer $W = W(k, r)$ such that for every $r$-coloring of $[W] = \{1, 2, \ldots, W\}$, some monochromatic arithmetic progression of $k$ terms must be formed.

As often happens, it turns out that it is easier to prove a somewhat stronger statement (which we take from [GR74]). First, we need some notation.

Let $[0, \ell]^m$ denote the set of $m$-tuples of nonnegative integers not exceeding $\ell$. Let us call two $m$-tuples $(x_1, \ldots, x_m)$, $(x_1', \ldots, x_m')$ *$\ell$-equivalent* if for some $i \geqslant 0$:

$$j < i \Rightarrow x_j = x_j' ,$$

$$x_i = x_i' = \ell ,$$

$$j > i \Rightarrow x_j < \ell ,\ x_j' < \ell .$$

If $i = 0$, then only the last condition applies. For any $\ell$, $m$, consider the statement:

$S(\ell, m)$: For any $r$, there exists $N = N(\ell, m, r)$ so that for any $r$-coloring $\chi$: $[N] \longrightarrow [r]$ there exist positive integers $a, d_1, \ldots, d_m$ such that $\chi(a + \sum_{i=1}^{m} x_i d_i)$ is constant on each $\ell$-equivalence class.

*Theorem 6.2* [GR74]

$$S(\ell, m) \text{ holds for all } \ell, m \geqslant 1 .$$

*Proof*: (i) $S(\ell, m)$ for some $m \geqslant 1 \Rightarrow S(\ell, m+1)$. For a fixed $r$, let $M = N(\ell, m, r)$, $M' = N(\ell, 1, r^M)$ and suppose $\chi$: $[MM'] \longrightarrow [r]$ is given. Define the induced coloring $\chi'$: $[M'] \longrightarrow [r^M]$ so that

$$\chi'(k) = \chi'(k') \textit{ iff } \chi(kM - j) = \chi(k'M - j) \text{ for } 0 \leqslant j < M .$$

By the induction hypothesis, there exist $a'$ and $d'$ such that $\chi'(a' + xd')$ is constant for $x \in [0, \ell - 1]$. Since $S(\ell, m)$ also applies to the interval $[(a' - 1)M + 1, a'M] := I$ then by the choice of $M$, there exist $a, d_1, \ldots, d_m$ with all sums $a + \sum_{i=1}^{m} x_i d_i$, $x_i \in [0, \ell]$, in $I$ and with $\chi(a + \sum_{i=1}^{m} x_i d_i)$ constant on $\ell$-equivalence classes. Set $d_i' = d_i$ for $i \in [m]$ and $d_{m+1}' = d'M$. Then $S(\ell, m+1)$ holds with these choices.

(ii) $S(\ell, m)$ for all $m \geqslant 1 \Rightarrow S(\ell + 1, 1)$. For a fixed $r$, let $\chi$: $[2N(\ell, r, r)] \longrightarrow [r]$ be arbitrarily given. Thus, there exist $a, d_1, \ldots, d_r$ such that for $x_i \in [0, \ell]$, $a + \sum_{i=1}^{r} x_i d_i$ is bounded above by $N(\ell, r, r)$ and $\chi(a + \sum_{i=1}^{r} x_i d_i)$ is constant on $\ell$-equivalence classes. By the pigeon-hole principle there exist $u, v \in [0, r]$ with $u < v$ such that

$$\chi(a + \sum_{i=1}^{u} \ell d_i) = \chi(a + \sum_{i=1}^{v} \ell d_i) .$$

Therefore,

$$\chi((a + \sum_{i=1}^{u} \ell d_i) + t(\sum_{i=u+1}^{v} d_i))$$

is constant for $t \in [0, \ell]$. This proves $S(\ell + 1, 1)$. Since $S(1, 1)$ clearly is true then the theorem holds by induction. ■

Of course, Theorem 6.1 is the special case $m = 1$ in Theorem 6.2.

The upper bound on $W(k, r)$ resulting from this proof is quite large. Essentially, it is given inductively by a function in two variables, and grows like the Ackermann function (cf. [Sp83]). In fact, no proof is known which yields an upper bound on $W(k, r)$ which is even primitive recursive! The same also applies to the special case $W(k) := W(k, 2)$.

On the other hand, the strongest lower bounds for $W(k)$ are much more modest, namely, just exponential in $k$ (cf. Theorems 6.3, 6.4). What the truth really is here represents a central open question in this whole area.

*Theorem 6.3* [Ber68]

If $p$ is prime, then

(6.1) $$W(p + 1) \geqslant p \cdot 2^p .$$

*Proof*: For simplicity, we only prove a slightly weaker result:

(6.2) $$W(p + 1) \geqslant p(2^p - 1) .$$

Let $GF(2^p)$ denote the finite field with $2^p$ elements, and fix a primitive element $\alpha \in GF(2^p)$. Let $v_1, \ldots, v_p$ be a basis for $GF(2^p)$ over $GF(2)$. For any integer $j$, set

$$\alpha^j = a_{1j} v_1 + a_{2j} v_2 + \ldots + a_{pj} v_p , \quad a_{ij} \in GF(2) .$$

Let

$$C_0 = \{j:\ a_{1j} = 0\,,\quad 1 \leqslant j \leqslant p(2^p - 1)\}\,,$$

$$C_1 = \{j:\ a_{1j} = 1\,,\quad 1 \leqslant j \leqslant p(2^p - 1)\}\,.$$

We claim that neither $C_0$ nor $C_1$ contains a $(p + 1)$-term arithmetic progression. Suppose, on the contrary, that $\{a, a + d, \ldots, a + pd\} \subseteq C_i$ for some $i$. Set $\beta = \alpha^a$, $\gamma = \alpha^d$. Since

$$1 \leqslant a < a + pd \leqslant p(2^p - 1)$$

then $d < 2^p - 1$ and so, since $\alpha$ is primitive, we have $\gamma \neq 1$. Therefore, $\beta, \beta\gamma, \ldots, \beta\gamma^{p-1}$ are all distinct and since $\beta\gamma^\ell = \alpha^{a+\ell d}$ then the elements $\beta, \beta\gamma, \ldots, \beta\gamma^p$ all have the same first coordinate, considered as vectors.

*Case 1.* $i = 0$. Then $\beta, \beta\gamma, \ldots, \beta\gamma^{p-1}$ are $p$ vectors in a $(p - 1)$-dimensional space (since the first coordinate is 0), and hence, they are dependent. Thus, there exist $a_0, a_1, \ldots, a_{p-1} \in GF(2)$, not all 0, such that

$$\sum_{i=0}^{p-1} a_i(\beta\gamma^i) = 0$$

and so,

$$\sum_{i=0}^{p-1} a_i\gamma^i = 0$$

which is impossible for $\gamma \in GF(2^p)$, $\gamma \neq 0, 1$ (cf. [MS78, Ch. 4, Th. 10]).

*Case 2.* $i = 1$. Thus, $\beta(\gamma - 1), \beta(\gamma^2 - 1), \ldots, \beta(\gamma^p - 1)$ belong to a $(p - 1)$-dimensional space, which implies

$$\sum_{i=0}^{p} a_i\beta(\gamma^i - 1) = 0$$

for $a_i \in GF(2)$ not all 0. Dividing by $\beta(\gamma - 1)$, we again see that $\gamma$ satisfies a polynomial of degree at most $p - 1$, a contradiction. ■

For general $k$, a slightly weaker lower bound is available (cf. [GRS80]).

*Theorem 6.4*

$$W(k) > \frac{2^k}{2e\,k}\,(1 + o(1))\,. \tag{6.3}$$

The proof is based on the following lemma, which is a consequence of Theorem 2.2.

*Lemma 6.5.* Let $A_1, A_2, \ldots, A_n$ be events with $Pr(A_i) \leqslant p$ for all $i$, and with a dependence graph having maximum degree $d$. Then

$$e\,p(d + 1) < 1 \;\Rightarrow\; Pr(\bigcap_i \bar{A}_i) > 0\,. \tag{6.4}$$

*Proof*: We apply Theorem 2.2 with $x_1 = \ldots = x_n = d/(d{+}1)$. The hypothesis of Theorem 2.2 then becomes

$$p < \frac{d^d}{(d + 1)^{d+1}}\,. \tag{6.5}$$

Since

$$p(d + 1)\left(1 + \frac{1}{d}\right)^d < e\,p(d + 1) < 1$$

then (6.5) is satisfied and thus, by Theorem 2.2, $Pr(\bigcap_i \bar{A}_i) > 0$. ■

*Proof of Theorem 6.4*: We red-blue color $[1, n]$ randomly so that each integer in $[1, n]$ is independently assigned the color red or blue with probability 1/2. To each $k$-term arithmetic progression $P$ associate the event $A_p$: "$P$ is monochromatic". Two vertices $P$ and $Q$ in the dependence graph $\Gamma$ form an edge if $P \cap Q \neq \varnothing$. The maximum degree $d$ in $\Gamma$ clearly satisfies $d \leqslant nk$. Thus, by Lemma 6.5, if $n \leqslant \frac{2^k}{2e\,k}(1 - o(1))$, then $Pr(\bigcap_P \bar{A}_P) > 0$. This implies that there are 2-colorings of $[1, n]$ having no monochromatic $k$-term arithmetic progressions and the claim is proved. ■

The estimation of $W(k)$ is closely related to the following problem:

Estimate

$$\nu_k(n) := \max\{|S|:\ S \subseteq [1, n],\ S \text{ contains no } k\text{-term arithmetic progression}\}\,.$$

The first major result for this problem was due to Roth [Ro53] who showed

(6.6) $$\nu_3(n) = O\left(\frac{n}{\log\log n}\right).$$

This was followed by the result of Szemerédi [Sz67] that $\nu_4(n) = o(n)$, and finally by the celebrated theorem of Szemerédi [Sz75], settling a \$1000 conjecture of Erdös and Turán:

*Theorem 6.5* [Sz75]

For all $k$,

(6.7) $$\nu_k(n) = o(n)\,.$$

(We will *not* present a proof of (6.7) here.) In this section we will restrict our discussion to the case $k = 3$. In this case, the best current bounds are given in the following result.

*Theorem 6.6*

(6.8) $$n \exp(-c_1 \sqrt{\log n}) < \nu_3(n) < c_2 n/(\log n)^{c_3}$$

for suitable positive constants $c_1, c_2, c_3$.

The lower bound is a classical result of Behrend [Beh46]; the upper bound is due to Szemerédi and Heath-Brown [H]. First, we give a proof of the lower bound.

For $d \geqslant 1$ and $x \leqslant n$, set

$$x = \sum_{j=0}^{k} x_j (2d+1)^j\,, \quad 0 \leqslant x_j \leqslant 2d\,.$$

Let

$$N(x_0, x_1, \ldots, x_k) := \Big(\sum_{i=0}^{k} x_i^2\Big)^{1/2}$$

and define

$$X_{n,d,s} := \{x:\ 1 \leqslant x \leqslant n\,,\ \ 0 \leqslant x_i \leqslant d \ \text{ for all } i\,, \ \text{ and }$$
$$N(x_0, x_1, \ldots, x_k) = \sqrt{s}\}\,.$$

*Claim*: $X_{n,d,s}$ contains no 3-term arithmetic progression.

*Proof*: Suppose that

$$x = \sum_{i=0}^{k} x_i (2d+1)^i ,$$

$$y = \sum_{i=0}^{k} y_i (2d+1)^i ,$$

$$z = \sum_{i=0}^{k} z_i (2d+1)^i$$

satisfy $x + y = 2z$ where $x \neq y$, $z \in X_{n,d,s}$. Since $x_i, y_i, z_i$ are all less than $d + 1$ then in fact we must have $x_i + y_i = 2z_i$ for all $i$. Furthermore,

$$N(x_0, x_1, \ldots, x_k) = N(y_0, y_1, \ldots, y_k) = N\left(\frac{x_0 + y_0}{2}, \ldots, \frac{x_k + y_k}{2}\right)$$

which implies

$$\sum_{i=0}^{k} (x_i - y_i)^2 = 0 ,$$

i.e., $x_i = y_i$ for all $i$, a contradiction.

For fixed $d$ we have

$$k \sim \frac{\log n}{\log(2d+1)}$$

and there are at most $d^2k$ possible values for $s$. The union of the $X_{n,d,s}$ over all $s$ contains all sums $\sum_i x_i(2d+1)^i$, $0 \leqslant x_i \leqslant d$, which are all at most $n$. There are essentially $n\,2^{-k}$ such integers. Thus, for some $s$

$$\nu_3(n) \geqslant |X_{n,d,s}| \geqslant \frac{n}{d^2 k 2^k} .$$

Setting $d = \exp(\sqrt{\log n})$ we deduce

$$|X_{n,d,s}| \geqslant n \exp(-c\sqrt{\log n})$$

for some $c > 0$, as required.

Instead of the upper bound in (6.8), we only show here that $\nu_3(n) = o(n)$. There are other relatively simple proofs of this fact (cf. [RS78], [G81]). The proof give here, based on ideas of Ruzsa and Szemerédi, is taken from [EFR86].

*Theorem 6.7*

(6.9) $$\nu_3(n) = o(n) .$$

*Proof*: Let $G = (V, E)$ denote a graph and let $A, B \subseteq V$ be a pair of disjoint non-empty subsets of $V$. The *density* of the pair $(A, B)$ is defined to be the ratio

$$d(A, B) := e(A, B)/|A||B|$$

where $e(A, B)$ denotes the number of edges $\{a, b\}$ with $a \in A$, $b \in B$. The pair $(A, B)$ is called $\epsilon$*-uniform* if for all $A' \subseteq A$, $B' \subseteq B$ with $|A'| > \epsilon|A|$, $|B'| > \epsilon|B|$ we have

$$|d(A', B') - d(A, B)| < \epsilon .$$

A partition $V = C_0 \cup C_1 \cup \ldots \cup C_k$ is called $\epsilon$*-uniform* if:

(1) $|C_0| < \epsilon|V|$;

(2) $|C_1| = |C_2| = \ldots = |C_k|$;

(3) All except at most $\epsilon \binom{k}{2}$ of the pairs $(C_i, C_j)$, $1 \leqslant i < j \leqslant k$, are $\epsilon$-uniform.

The following fundamental result (which we state here without proof) is due to Szemerédi.

*Lemma 6.8.* (Regularity Lemma [Sz76]). For every $\epsilon > 0$ and positive integer $\ell$, there exist positive integers $n_0(\epsilon, \ell)$ and $k_0(\epsilon, \ell)$ such that every graph with at least $n_0(\epsilon, \ell)$ vertices has an $\epsilon$-uniform partition into $k$ classes, where $k$ is some integer satisfying $\ell < k < k_0(\epsilon, \ell)$.

Now, let $A \subseteq [1, n]$ and let $X, Y, Z$ be three disjoint copies of the interval $[1, 3n]$. Consider the set $S$ of all triples $\{x, y, z\}$, $x \in X, y \in Y, z \in Z$ such that

$$y - x = z - y = \frac{z - x}{2} \in A\,. \tag{6.10}$$

Further, let $G$ be the graph consisting of all pairs contained in the triples of $S$. Thus, $G$ has $9n$ vertices, $|E(G)| \geqslant 3|A|n$ edges, and $E(G)$ can be decomposed $\frac{1}{3}|E(G)|$ edge-disjoint triangles $\{x, y, z\}$, which we call "ordinary" triangles.

*Claim 6.9.* [RS78] If $G$ contains a triangle which is not ordinary, then $A$ contains a 3-term arithmetic progression.

To see this, let $x', y', z'$ be such a triangle where $y' - x' \neq z' - y'$. Thus, for $a := y' - x'$ and $b := z' - y'$, we have $a \in A$, $b \in A$, and $\frac{a+b}{2} = \frac{z' - x'}{2} \in A$, which forms the required 3-term arithmetic progression.

Suppose now that $|A| = \alpha n$ for a fixed positive constant $\alpha$ independent of $n$. We will show that for $n$ sufficiently large, $A$ must contain a 3-term arithmetic progression. Set $m = 9n$ and set

$$|E(G)| = \beta\binom{m}{2} > 3\alpha n^2$$

where $\beta$ is a fixed positive constant independent of $n$. Further, set $\epsilon = \beta/15$ and $\ell = \lceil \epsilon^{-1} \rceil$. We now apply the Regularity Lemma to $G$ with these choices (and $n > n_0(\epsilon, \ell)$). The number of edges not contained in pairs with density at least $\beta/6$ is at most

$$k\binom{m/k}{2} + \frac{\beta}{6}\binom{k}{2}\left(\frac{m}{k}\right)^2 + \epsilon\binom{k}{2}\left(\frac{m}{k}\right)^2 + \epsilon m^2 < \frac{\beta}{3}\binom{m}{2}\,.$$

After the deletion of these edges we obtain a graph $G'$ which still contains a triangle $T$ (there were $\frac{\beta}{3}\binom{m}{2}$ edge-disjoint triangles in $G$). Moreover, all three edges of this triangle are contained in pairs which are $\epsilon$-uniform and which have density at least $\beta/6$. Let $C_p$, $C_q$ and $C_r$ be the three partition classes containing the three endpoints of $T$.

*Claim 6.10.* If all three pairs $(C_p, C_q)$, $(C_p, C_r)$ and $(C_q, C_r)$ are $\epsilon$-uniform with density at least $\beta/6$ then there is a vertex $x \in C_r$ which is contained in at least $\left(\frac{\beta}{10}\right)^3 |C_p||C_q|$ triangles.

*Proof*: If both $(C_p, C_r)$ and $(C_q, C_r)$ are $\epsilon$-uniform then we can find at least $(1 - 2\epsilon)|C_r|$ vertices $x \in C_r$ which are joined to at least $\left(\frac{\beta}{6} - \epsilon\right)|C_i|$ vertices of $C_i$, for $i = p$ and $i = q$. Fix one such vertex $x$ and let $N_x^i$ be the set of neighbors of $x$ in $C_i$. Since $\frac{\beta}{6} - \epsilon = \frac{\beta}{10} > \epsilon$, we find there are at least $\left(\frac{\beta}{10}\right)^3 |C_p||C_q|$ edges joining vertices of $N_x^p$ and $N_x^q$. Each such edge clearly corresponds to a triangle containing $x$. ■

To complete the proof of Theorem 6.7, note that by Claim 6.10, for $m$ sufficiently large (and $k$ fixed) there are at least

$$(6.11) \qquad \left(\frac{\beta}{10}\right)^3 |C_p||C_q| > |C_p| = |C_q|$$

triangles containing the vertex $x$. Since no two ordinary triangles in $G$ share two vertices then there are at most $|C_p| = |C_q|$ ordinary triangles containing the vertex $x$. Thus, by (6.11), $G$ contains a triangle which is not ordinary, and so by Claim 6.9, $A$ contains a 3-term arithmetic progression. This completes the proof of Theorem 6.7. ■

We conclude this section with several remarks. To begin with, as an indication of the extent of our ignorance on the true order of growth of the van der Waerden function $W(k)$, the first author has made for some time the following offer.

*Conjecture* (\$1000) For all $k$,

$$W(k) \leqslant \left.2^{2^{2^{\cdot^{\cdot^{2}}}}}\right\} k\,.$$

The known values are:

$$W(2) = 3\,, \quad W(3) = 9\,, \quad W(4) = 35\,, \quad W(5) = 178\,.$$

An interesting variation which was considered in [G83] is the following. Define $W^*(k)$ to be the least integer such that there exists a set $X(k) \subseteq \{1, 2, \ldots\}$ with $|X(k)| = W^*(k)$ so that any 2-coloring of $X(k)$ always forms a monochromatic $k$-term arithmetic progression.

*Problem 6.11*

$$\text{Does } W^*(k)/W(k) \longrightarrow 0 \text{ as } k \longrightarrow \infty\,?$$

It is known [G83] that $W^*(2) = W(2)$, $W^*(3) = W(3)$ but $W^*(4) \leqslant 27 < 35 = W(4)$.

## 7. Concluding remarks

This paper has dealt almost exclusively with asymptotic bounds for various results of Ramsey type. The *exact* values for the associated functions are invariably much more difficult to obtain. As an indication of this difficulty, we list in Table 1, all known (non-trivial) values of $r(k, \ell)$, $k \leqslant \ell$, together with the best bounds currently available for several other values of $r(k, \ell)$. It would appear, for example, that the determination of $r(5, 5)$ will require some significant new ideas. The reader is referred to [CG83] or [RK] for a fuller discussion.

| $k \backslash \ell$ | 3 | 4 | 5 | 6 | 7 | 8 | 9 | 10 |
|---|---|---|---|---|---|---|---|---|
| 3 | 6 | 9 | 14 | 18 | 23 | 28–29 | 36 | 39–44 |
| 4 | – | 18 | 25–28 | 39–44 | | | | |
| 5 | – | – | 42–55 | 57–94 | | | | |
| 6 | – | – | – | 102–169 | | | | |

Small values of $r(k, \ell)$

**Table 1**

In the other direction, the best upper bound we currently have for the van der Waerden function $W(k)$ grows like the Ackermann function. While this is not known to be the true order of growth of $W(k)$ (and, in the opinion of many combinatorialists, is a gross over-estimate), it turns out that there are in fact a number of natural problems of Ramsey type which *do* have functions which grow this rapidly, and indeed, *much* more rapidly. The earliest example of this phenomenon was exhibited in the celebrated result of Paris and Harrington [PH77]. To state their result, we need one definition. Let us call a finite set $S$ of positive integers *large* if $\min(S) > |S|$. Consider the following statement:

(7.1) For all $k$ and $t$, there exists a least number $PH(k, t)$ so that if $n \geqslant PH(k, t)$ then in any $t$-coloring of the $k$-element subsets of $[1, n]$ there must exist a monochromatic *large* set $B \subseteq [k+1, n]$

The truth of (7.1) follows easily from the infinite form of Ramsey's theorem. What was unexpected was that while (7.1) is a perfectly well-defined statement in Peano Arithmetic (PA) (that first order theory of numbers which includes the basically finitistic methods of number theory), it is in fact *unprovable* in PA. One way of proving this, as pointed out by Ketonen and Solovay [KS81], is to show that the function $PH(k, t)$ grows so rapidly that it cannot even be defined in PA. A nice-description of this work is given in [Sm80], [Sm82] and [Sp83].

In some sense, even more striking is a very recent result of Friedman, which can be considered as a finite form of a well known result of Kruskal [Kr60] (which asserts that the set of finite trees is well-quasi-ordered under homeomorphic embedding). This finite form is given by the following statement:

(7.2) For each $k > 1$, there is a least number $F(k)$ such that if $n \geqslant F(k)$ and $T_1, T_2, \ldots, T_n$ is a sequence of trees with $T_i$ having at most $k + i$ vertices then there exist $i < j$ such that $T_i$ is homeomorphically embeddable in $T_j$.

Although (7.2) only deals with finite sets of finite objects, any proof of (7.2) must, in a certain precise sense, invoke the concept of uncountability. Again, this is due from one point of view to the extremely rapid growth rate of the function $F(k)$. A full account of these fascinating developments can be found in [Sm82], [Sp83] and [NT87].

*Acknowledgements.* The authors gratefully acknowledge the useful comments of P. Erdös and Z. Füredi given during the preparation of this paper.

**REFERENCES**


[A72] H. L. Abbott, "Lower bounds for some Ramsey numbers", Disc. Math. 2 (1972), 289-293.

[AKS80] M. Ajtai, J. Komlós, and E. Szemerédi, "A note on Ramsey numbers", J. Comb. Th. (A) 29 (1980), 354-360.

[AKS81a] ______________, "A dense infinite Sidon sequence", Eur. J. Comb. 2 (1981), 1-15.

[AKS81b] ______________, "On Turán's theorem for sparse graphs", Combinatorica 1 (1981), 313-317.

[AC] N. Alon and F. R. K. Chung, "Explicit constructions of linear-sized tolerant networks", (preprint), 1986.

[Bec83a] J. Beck, "On size Ramsey numbers of paths, trees, and circuits I", J. Graph. Th. 7 (1983), 115-129.

[Bec83b] ______________, "An upper bound for diagonal Ramsey numbers", Studia Sci. Math. Hung. 18 (1983), 401-406.

[Beh46] F. A. Behrend, "On sets of integers which contain no three in arithmetic progression", Proc. Nat. Acad. Sci. 23 (1946), 331-332.

[Ber68] E. Berlekamp, "A construction for partitions avoiding long arithmetic progressions", Canad. Math. Bull. 11 (1968), 409-414.

[BT81] B. Bollobás and A. G. Thomason, "Graphs which contain all small graphs", Eur. J. Comb. 2 (1981), 13-15.

[Bu74] S. A. Burr, "Generalized Ramsey theory for graphs — a survey" in Graphs and Combinatorics, Springer-Verlag, Berlin, 1974, 52-75.

[Bu70] ______________, "A survey of noncomplete Ramsey theory for graphs", Ann. New York Acad. Sci. 328 (1979), 58-75.

[BE75] S. A. Burr and P. Erdös, "On the magnitude of generalized Ramsey numbers for graphs", Colloq. Math. Soc. J. Bolyai 10, North-Holland, 1975, 214-240.

[BE76] ______________, "Extremal Ramsey theory for graphs", Utilitas Math. 9 (1976), 247-258.

[Ch73] F. R. K. Chung, "On the Ramsey numbers $N(3,3,\ldots,3;2)$, Disc. Math. 5 (1973), 317-321.

[Ch74] ______________, "On triangular and cyclic Ramsey numbers with $k$ colors", Lecture Notes in Math. No. 406 (1974), 236-242, Springer Verlag, New York.

[CG75] F. R. K. Chung and R. L. Graham, "On multicolor Ramsey numbers for complete bipartite graphs", J. Comb. Th. (A) 18 (1975), 164-169.

[CG83] F. R. K. Chung and C. M. Grinstead, "A survey of bounds for classical Ramsey numbers", J. Graph Th. 8 (1983), 25-37.

[C77] V. Chvátal, "Tree-complete graph Ramsey numbers", J. Graph Th. 1 (1977), 93.

[CH72] V. Chvátal, and F. Harary, "Generalized Ramsey theory for graphs III: Small off-diagonal numbers", Pacific J. Math. 41 (1972), 335-345.

[CRST83] C. Chvátal, V. Rödl, E. Szemerédi and W. T. Trotter, Jr., "The Ramsey number of a graph with bounded maximum degree", J. Comb. Th. (B) 34 (1983), 239-243.

[D75] W. Deuber, "A generalization of Ramsey's theorem", in Infinite and Finite Sets, A. Hajnal, R. Rado and V. T. Sós, eds., Colloq. Math. Soc. J. Bolyai 10, North-Holland, Amsterdam, 1975, 323-332.

[E47] P. Erdös, "Some remarks on the theory of graphs", Bull. Amer. Math. Soc. 53 (1947), 292-294.

[E57] ______________, "Remarks on a theorem of Ramsey", Bull. Res. Council Israel Sect. F7 (1957), 21-24.

[E61] ______________, "Graph theory and probability II", Canad. J. Math. 13 (1961), 346-352.

[E65] ______________, "On extremal problems of graphs and generalized graphs", Israel J. Math. 2 (1965), 189-190.

[E66] ______________, "On the construction of certain graphs", J. Comb. Th. 1 (1966), 149-153.

[E75] ______________, "Problems and results on finite and infinite graphs", in Recent Advances in Graph Theory, M. Fiedler ed., Academia Praha, 1975, 183-192.

[EFR86] P. Erdös, P. Frankl and V. Rödl, "The asymptotic number of graphs not containing a fixed subgraph and a problem for hypergraphs having no exponent", Graphs and Combinatorics 2 (1986), 113-121.

[EG73] P. Erdös and R. L. Graham, "On partition theorems for finite graphs", Colloq. Math. Soc. J. Bolyai 10 (1973), 515-527.

[EH72] P. Erdös and A. Hajnal, "On Ramsey-like theorems. Problems and results", Proc. Conf. Combinatorial Math, Math. Inst. Oxford (1972). Inst. Math. Appl. Southend-in Sea, 1972, 123-140.

[EH] ______________, unpublished.

[EHMR84] P. Erdös, A. Hajnal, A. Máté and R. Rado, Combinatorial Set Theory: Partition relations for cardinals, North-Holland, Amsterdam, 1984.

[EHP75] P. Erdös, A. Hajnal and L. Pósa, Strong embeddings of graphs into colored graphs", in Infinite and Finite sets, A. Hajnal, R. Rado and V. T. Sós, eds., Colloq. Math. Soc. J. Bolyai 10, North-Holland, Amsterdam, 1975, 585-595.

[EHR65] P. Erdös, A. Hajnal and R. Rado, "Partition relations for cardinal numbers", Acta Math. Acad. Sci. Hungar. 16 (1965), 93-196.

[EL75] P. Erdös and L. Lovász, "Problems and results on 3-chromatic hypergraphs and some related questions", in Infinite and Finite Sets, A. Hajnal, R. Rado and V. T. Sós, eds., North-Holland, Amsterdam, 1975, 609-628.

[EM81] P. Erdös and G. Mills, "Some bounds for the Ramsey-Paris-Harrington numbers", J. Comb. Th. (A) 30 (1981), 53-70.

[ER52] P. Erdös and R. Rado, "Combinatorial theorems on classifications of subsets of a given set", Proc. London Math. Soc. 2 (1952), 417-439.

[ES74] P. Erdös and J. H. Spencer, Probabilistic Methods in Combinatorics, Akademiai Kiadó, Budapest, 1974.

[ES35] P. Erdös and G. Szekeres, "A combinatorial problem in geometry", Compositio Math. 2 (1935), 463-470.

[Fr77] P. Frankl, "A constructive lower bound for Ramsey numbers", Ars. Comb. 3 (1977), 297-302.

[FW81] P. Frankl and R. M. Wilson, "Intersection theorems will geometric consequences", Combinatorica 1 (1981), 357-368.

[Fr79] H. Frederickson, "Schur numbers and the Ramsey numbers $N(3, \ldots, 3;2)$, J. Comb. Th. (A) 27 (1979), 376-377.

[FP] J. Friedman and N. Pippenger, "Expanding graphs contain all small trees", (preprint), 1986.

[Fu85] Z. Füredi, "A Ramsey Sperner theorem", Graphs and Combinatorics 1 (1985), 51-56.

[Go59] A. W. Goodman, "On sets of acquaintances and strangers at any party", Amer. Math. Monthly 66 (1959), 778-783.

[G81] R. L. Graham, Rudiments of Ramsey Theory, CBMS Regional Conference in Math. no. 45, Amer. Math. Soc., Providence, 1981.

[G83] ______________, "Recent developments in Ramsey theory", Proc. Int'l. Cong. Math. Warsaw, 1983, 1555-1567.

[GR71a] R. L. Graham and B. L. Rothschild, "Ramsey's theorem for $n$-parameter sets", Trans. Amer. Math, Soc. 159 (1971), 257-292.

[GR71b] ______________, "A survey of finite Ramsey theorems", Proc. 2nd Louisiana Conf. on Combinatorics, Graph Theory and Computing (1971), 21-40.

[GR74] ______________, "A short proof of van der Waerden's theorem on arithmetic progressions", Proc. Amer. Math. Soc. 42 (1974), 356-386.

[GRS80] R. L. Graham, B. L. Rothschild and J. H. Spencer, Ramsey Theory, John Wiley and Sons, New York, 1980.

[GS71] R. L. Graham and J. H. Spencer, "A constructive solution to a tournament problem", Canad. Math. Bull. 14 (1971), 45-48.

[GY68] J. E. Graver and J. Yackel, "Some graph theoretic results associated with Ramsey's theorem", J. Comb. Th. 4 (1968), 125-175.

[GG55] R. E. Greenwood and A. M. Gleason, "Combinatorial relations and chromatic graphs", Canad. J. Math. 7 (1955), 1-7.

[H] D. R. Heath-Brown, "Integer sets containing no arithmetic progressions", (preprint), 1986.

[I74] R. W. Irving "Generalized Ramsey numbers for small graphs", Disc. Math. 9 (1974), 251-264.

[KS81] J. Ketonen and R. Solovay, "Rapidly growing Ramsey functions", Ann. Math. 113 (1981), 267-314.

[Kr60] J. B. Kruskal, "Well-quasi-ordering, the tree theorem, and Vázonyi's conjecture", Trans. Amer. Math. Soc. 95 (1960), 210-225.

[LSW] E. Levine, J. H. Spencer and J. Winn, personal communication.

[L79] L. Lovász, Combinatorial Problems and Exercises, North-Holland, Amsterdam, 1979.

[LRS] A. Lubotzky, R. Phillips and P. Sarnak, "Ramanujan graphs", (preprint), 1986.

[MS78] F. J. MacWilliams and N. J. A. Sloane, The Theory of Error-Correcting Codes, North-Holland, Amsterdam, 1978.

[Ma87] R. Mathon, "Lower bounds for Ramsey numbers and association schemes", J. Comb. Th. (B) 42 (1987), 122-127.

[M85] G. Mills, "Ramsey-Paris-Harrington numbers for graphs", J. Comb. Th. (A) 38 (1985), 30-37.

[NR78] J. Nešetřil and V. Rödl, "Partition (Ramsey) theory - a survey", Colloq. Math. Soc. J. Bolyai 18, North-Holland, Amsterdam 1978, 754-792.

[NR79] ______________, "Partition theory and its application", in Surveys in Combinatorics, London Math. Soc. Lecture Note Series no. 38, Cambridge Univ. Press, London 1979, 96-149.

[NT87] J. Nešetřil and R. Thomas, "WQO, long games and a combinatorial study of unprovability", in Logic and Combinatorics, Contemporary Math., Amer. Math. Soc., Providence, 1987.

[PH77] J. Paris and L. Harrington, "A mathematical incompleteness in Peano Arithmetic", in Handbook of Mathematical Logic, J. Barwise, ed., North-Holland, Amsterdam, 1977, 1133-1142.

[Ra33] R. Rado, "Studien zur Kombinatorik", Math. Z. 36 (1933), 425-480.

[RK] S. P. Radziszowski and D. L. Kreher, "Search algorithm for Ramsey graphs by union of group orbits", (preprint), 1986.

[R30] F. P. Ramsey, "On a problem of formal logic", Proc. London Math. Soc. 30 (1930), 264-286.

[RW75] D. K. Ray-Chaudhuri and R. M. Wilson, "On $t$-designs", Osaka J. Math. 12 (1975), 735-744.

[R73] V. Rödl, "The dimension of a graph and generalized Ramsey theorems", Thesis, Charles Univ, Praha, 1973.

[R82] ______________, "Note on finite Boolean Algebras", Acta Polytechnica (1982), 47-49.

[R] ______________, "Upper bounds on Ramsey numbers $R(k,\ell)$", (unpublished manuscript), 1986.

[RS] V. Rödl and E. Szemerédi, unpublished.

[Ro53] K. Roth, "On certain sets of integers", J. London Math. Soc. 28 (1953), 104-109.

[RS78] I. Z. Ruzsa and E. Szemerédi, "Triple systems with no six points carrying three triangles", Colloq. Math. Soc. J. Bolyai 18 (1978), 939-945.

[Sa68] J. H. Sanders, "A generalization of Schur's theorem", Doctoral Dissertation, Yale Univ., 1968.

[Sc16] I. Schur, "Uber die Kongruenz $x^m + y^m \equiv z^m \ (mod\, p)$", Jber. Deutsche Math-Verein. 25 (1916), 114-116.

[Sh83] J. B. Shearer, "A note on the independence number of a triangle-free graph", Disc. Math. 46 (1983), 83-87.

[Sh86] ______________, "Lower bounds for small diagonal Ramsey numbers", J. Comb. Th. (A) 42 (1986), 302-304.

[Sm80] C. Smoryński, "Some rapidly growing functions", Math. Intelligencer 2 (1980), 149-154.

[Sm82] ______________, "The varieties of arboreal experience", Math. Intelligencer 4 (1982), 182-189.

[Sp75] J. H. Spencer, "Ramsey's theorem – a new lower bound", J. Comb. Th. (A) 18 (1975), 108-115.

[Sp77] ______________, "Asymptotic lower bounds for Ramsey functions", Disc. Math. 20 (1977), 69-76.

[Sp83] ______________, "Large numbers and unprovable theorems", Amer. Math. Monthly 90 (1983), 669-675.

[Sz75] E. Szemerédi, "On sets of integers containing no *k* elements in arithmetic progression", Acta Arith. 27 (1975), 199-245.

[Sz76] ______________, "Regular partitions of graphs", Proc. Colloq. Int. CNRS, CNRS, Paris, 1976, 399-401.

[T81] A. D. Taylor, "Bounds for the disjoint unions theorem", J. Comb. Th. (A) 39 (1981), 339-344.

[W27] B. L. van der Waerden, "Beweis einer Baudetschen Vermutung", Nieuw Arch. Wisk. 15 (1927), 212-216.

[Y72] J. Yackel, "Inequalities and asymptotic bounds for Ramsey numbers", J. Comb. Th. 13 (1972), 56-58.

# ALMOST DISJOINT SETS

E.C. Milner
University of Calgary, Calgary, Alberta, T2N-1N4, Canada.

K.Prikry
University of Minnesota, Minneapolis, Minnesota, 55455, U.S.A.

## 1 BASIC RESULTS

Two sets A, B are said to be almost disjoint if

$$|A\cap B| < \min\{|A|, |B|\},$$

and a family, $\mathcal{F}$, of sets is called an almost disjoint family if its members are pairwise almost disjoint. This simple concept, which seems to have first been considered by SIERPINSKI, is quite a useful one and it also gives rise to some interesting questions.

We shall use standard set-theoretic notation so that, for example, an ordinal $\alpha$ is the set $\{\beta : \beta < \alpha\}$ of all smaller ordinals, $\mathrm{cf}\kappa$ is the cofinality and $\kappa^+$ is the successor cardinal of the cardinal $\kappa$. Greek letters are used to denote ordinal numbers and, in particular, the letters $\kappa$, $\lambda$, $\mu$ will always denote infinite cardinals. We define $[S]^\lambda = \{X \subseteq S: |X| = \lambda\}$, $[S]^{<\lambda} = \{X \subseteq S: |X| < \lambda\}$. A tree is a partially ordered set $(T,\le)$ such that the set of predecessors of any element t, $T(<t) = \{x \in T: x < t\}$, is well ordered by $<$. The height of an element $t \in T$ is the order type of $T(<t)$; the $\alpha$-th level, $T_\alpha$, of the tree is the set of elements at height $\alpha$. The height of the tree , $h(T)$, is the least $\alpha$ such that $T_\alpha = \emptyset$. A branch of $(T,\le)$ is a maximal chain in T, and a $\mu$-branch is a branch of order type $\mu$. A family of sets, $\mathcal{F}$, is called a $(\kappa,\lambda,\mu)$-family if (i) $|\mathcal{F}| = \kappa$, (ii) $\mathcal{F} \subseteq \mathcal{P}(\lambda)$, and (iii) $|A| = \mu$ for each $A \in \mathcal{F}$. A $(\kappa,\lambda,-)$-family (respectively a $(\kappa,-,\mu)$-family) is one which satisfies (i) & (ii) (respectively (i) & (iii)).

SIERPINSKI [1928] was the first to observe that there is an almost disjoint $(\kappa,\lambda,-)$-family with $\kappa > \lambda$, a result which he described as "un peu paradoxal", presumably in view of the fact that there can be at most $\lambda$ pairwise disjoint subsets of $\lambda$. This naturally raises the question how large can an almost disjoint family in $\mathcal{P}(\lambda)$ be, and this problem was investigated rather more thoroughly by TARSKI [1928,1929] and

BAUMGARTNER [1976]. The simplest non-trivial example of an almost disjoint family of sets is the set of $\mu$-branches in a tree of height $\mu$.

THEOREM 1.1. Let $\lambda$, $\mu$, $\rho$ be cardinals ($\rho$ may be finite), let $\tau$ be either $\mu$ or cf$\mu$, and suppose that $\mu$ is minimal such that $\rho^{\mu} > \lambda$. Then there is a tree of cardinality $\leq \lambda$ with height $\tau$ and having $\rho^{\mu}$ $\tau$-branches.

COROLLARY 1.2. If $\lambda$, $\mu$, $\rho$, $\tau$ are as in Theorem 1.1, then there is an almost disjoint $(\rho^{\mu},\lambda,\tau)$-family. In particular, there is an almost disjoint $(2^{\omega},\omega,\omega)$-family.

PROOF. If $\tau = \mu$, consider the tree, T, of functions f: $\alpha \to \rho$ $(\alpha < \mu)$, ordered by inclusion; this has height $\mu$, cardinality $\leq \lambda$, and has $\rho^{\mu}$ $\mu$-branches. If $\tau = \text{cf}\mu < \mu$, choose cardinals $\mu_{\alpha}$ $(\alpha < \text{cf}\mu)$ with supremum $\mu$ and consider instead the tree consisting of functions f: $\mu_{\alpha} \to \rho$ $(\alpha < \text{cf}\mu)$; this has height cf$\mu$, cardinality $\leq \lambda$, and has $\rho^{\mu}$ cf$\mu$-branches. □

It follows from Corollary 1.2 that, for any infinite cardinal $\lambda$, there is an almost disjoint $(\lambda^{+},\lambda,-)$-family. In fact, if $\lambda$ is a strong limit cardinal there is a tree of height and cardinality $\lambda$ having $2^{\lambda}$ $\tau$-branches where $\tau = \lambda$ or cf$\lambda$ and hence there is an almost disjoint $(2^{\lambda},\lambda,\tau)$-family. If we assume the generalized continuum hypothesis (GCH), then $2^{<\lambda} = \lambda$ for any $\lambda$ and so by Corollary 1.2 that there is an almost disjoint $(\lambda^{+},\lambda,\lambda)$-family and an almost disjoint $(\lambda^{+},\lambda,\text{cf}\lambda)$-family. TARSKI [1928,1929] proved the more general result that, assuming GCH, there is an almost disjoint $(\lambda^{+},\lambda,\mu)$-family for any cardinal $\mu \leq \lambda$ such that cf$\mu$ = cf$\lambda$. However, GCH is not really needed to prove this (see Theorem 1.3). Before proving this result let us observe that Corollary 1.2 gives a little more in the case when $\mu = \lambda$ or cf$\lambda$. For example, if we assume just CH, i.e. $2^{\omega} = \omega_1$, but allow $2^{\omega_1}$ to be arbitrarily large, then by the corollary there is an almost disjoint $(2^{\omega_1},\omega_1,\omega_1)$-family.

THEOREM 1.3. If $\mu \leq \lambda$ and cf$\mu$ = cf$\lambda$, then there is an almost disjoint $(\lambda^{+},\lambda,\mu)$-family.

PROOF. Suppose first that $\mu$ = cf$\lambda$. It is enough to show that, if $\mathcal{F} = \{A_{\alpha}\colon \alpha < \lambda\}$ is an almost disjoint $(\lambda,\lambda,\mu)$-family, then $\mathcal{F}$ is not maximal.

There are sets $X_\alpha \subseteq \lambda$ $(\alpha < cf\lambda)$ such that $|X_\alpha| < \lambda$ and $\cup\{X_\alpha: \alpha < cf\lambda\} = \lambda$. For $\alpha < cf\lambda$ choose $y_\alpha \in \lambda \setminus Z_\alpha$, where

$$Z_\alpha = \cup\{A_\beta: \beta \in \cup\{X_\gamma: \gamma < \alpha\}\}\cup\{y_\beta:\beta < \alpha\}.$$

This choice is possible for, if $\lambda$ is regular $|A_\alpha \setminus Z_\alpha| < \lambda$, and if $\lambda$ is singular $|Z_\alpha| < \lambda$. The set $Y = \cup\{y_\alpha: \alpha < cf\lambda\}$ is almost disjoint from each $A \in \mathcal{F}$ and so $\mathcal{F}$ is not maximal.

Now suppose that $\mu > cf\lambda$ is singular. Choose increasing cardinals $\mu_\beta < \mu$ $(\beta < cf\lambda)$ with supremum $\mu$. For each $\alpha < \lambda$ with $cf\alpha = cf\lambda$, choose an increasing continuous sequence of ordinals $\xi(\alpha,\beta)$ $(\beta < cf\lambda)$ cofinal in $\alpha$, and $\xi(\alpha,0) = 0$. Then, for every $\xi < \alpha$, there is $\beta = \beta(\alpha,\xi) < cf\lambda$ such that $\xi(\alpha,\beta) \leq \xi < \xi(\alpha, \beta+1)$. For $\xi < \alpha < \lambda$ and $cf\alpha = cf\lambda$, choose pairwise disjoint subsets $X(\alpha,\xi)$ of $\lambda$ such that $|X(\alpha,\xi)| = \mu_{\beta(\alpha,\xi)}$. By the first part there is an almost disjoint $(\lambda^+,\lambda,cf\lambda)$-family, $\mathcal{F}$, and we may assume that each $A \in \mathcal{F}$ has order type $cf\lambda$ (in the natural ordering of $\lambda$). For $A \in \mathcal{F}$ put $f(A) = \cup\{X(\alpha,\xi): \xi \in A\}$, where $\alpha = \sup A$. Then the family $\mathcal{F}'$ $= \{f(A): A \in \mathcal{F}\}$ is an almost disjoint $(\lambda^+,\lambda,\mu)$-family. □

Note that GCH was not used in the above proof. If we do assume GCH, then the result follows almost immediately from Theorem 1.1. (There is a tree $(T,\leq)$ of cardinality $\lambda$ and height $cf\lambda$ which has $\lambda^+$ branches; replace each element $t \in T$ at height $\alpha$ $(\alpha < cf\lambda)$ by a set $X(t)$ of cardinality $\mu_\alpha$, the sets $X(t)$ being pairwise disjoint, and consider the family of sets $f(B) = \cup\{X(t): t \in B\}$ ($B$ a branch of $T$).) In view of this the reader may wonder if the corresponding stronger statement about branches in trees can be proved by adapting the argument used in the proof of Theorem 1.3. For example, is it true that $\mathcal{S}$ : <u>there is a tree of height and cardinality $\lambda$ having at least $\lambda^+$ $\lambda$-branches.</u> However, the situation is quite different. MITCHELL [1972] has shown that it is not possible to prove $\mathcal{S}$ without GCH.

In this connection, let us note the following theorem (and corollaries) of BAUMGARTNER [1976].

THEOREM 1.4. <u>If $\mathcal{F}$ is a $(\kappa,\lambda,\mu)$-family, and if</u> $|\{A \in \mathcal{F}: |A\cap\alpha| = \mu\}| < \kappa$ <u>whenever</u> $\alpha < \lambda$, <u>then</u> $cf\lambda = cf\kappa$ <u>or</u> $cf\mu$.

<u>PROOF</u>. Suppose false. Then for each $A \in \mathcal{F}$, since $cf\lambda \neq cf\mu$, there is $\alpha_A < \lambda$ such that $|A\cap\alpha| = \mu$. Also, since $cf\lambda \neq cf\kappa$, there is $\alpha < \lambda$ such that $\alpha_A < \alpha$ for $\kappa$ different sets $A \in \mathcal{F}$. But this contradicts the hypothesis. □

COROLLARY 1.5. Let $\lambda$ be the least cardinal such that there is a tree of cardinality $\lambda$ and height $\mu$ which has at least $\kappa$ $\mu$-branches. Then $cf\lambda = cf\kappa$ or $cf\lambda = cf\mu$.

PROOF. Let $(T,\leq)$ be a tree of cardinality $\lambda$ and height $\mu$ having at least $\kappa$ $\mu$-branches. We may assume that $T = \lambda$. If $cf\lambda \neq cf\kappa$ or $cf\mu$, then by Theorem 1.4 there is $\alpha < \lambda$ such that $|B\cap\alpha| = \mu$ for $\kappa$ different branches B of T. But then the subtree on $\alpha$ has height $\mu$ and has at least $\kappa$ $\mu$-branches, and this contradicts the minimality of $\lambda$. □

Also, essentially by the same proof, we have the following.

COROLLARY 1.6. If $\lambda$ is the least cardinal such that there is an almost disjoint $(\kappa,\lambda,\mu)$-family, then $cf\lambda = cf\kappa$ or $cf\mu$.

TARSKI [1928] introduced another parameter to measure the 'almost disjointedness' of a family. The degree of disjunction of $\mathcal{F}$ is

$$\delta(\mathcal{F}) = \min\{\nu: |A\cap B| < \nu \text{ whenever } A,B \text{ are distinct members of } \mathcal{F}\}.$$

Thus a $(\kappa,\lambda,\mu)$-family $\mathcal{F}$ is almost disjoint if $\delta(\mathcal{F}) \leq \mu$. If $\delta(\mathcal{F}) < \mu$ then $\mathcal{F}$ is said to be strongly almost disjoint. We define a $(\kappa,\lambda,\mu,\nu)$-family to be a $(\kappa,\lambda,\mu)$-family with $\delta(\mathcal{F}) \leq \nu$. TARSKI [1928] proved:

THEOREM 1.7. If $\mathcal{F}$ is a $(\kappa,\lambda,-)$-family with $\delta(\mathcal{F}) \leq \nu$, then $\kappa \leq \lambda^{\nu}$.

COROLLARY 1.8. There is no $(\kappa,\lambda,\mu,\nu)$-family with $\kappa > \lambda^{\nu}$.

PROOF. For each $A \in \mathcal{F}$ choose $f(A) \in [A]^{\nu}$ if $|A| > \nu$, and put $f(A) = A$ if $|A| \leq \nu$. Then $f:\mathcal{F} \to [\lambda]^{\nu}$ is one-one. □

We have seen (without assuming GCH, Theorem 1.3) that, for any infinite cardinal $\lambda$, there is a $(\lambda^{+},\lambda,\mu,\mu)$-family if $\mu \leq \lambda$ and $cf\mu = cf\lambda$. TARSKI [1929] also showed, assuming GCH, that this condition on $\mu$ is necessary. The following theorem is a strengthening of this due to BAUMGARTNER [1976].

THEOREM 1.9. Assume that $\lambda\geq\mu\geq\nu$. Then (i) there is a $(\lambda^{+},\lambda,\mu,\mu)$-family if $cf\mu = cf\lambda$; (ii) if $2^{\tau} \leq \lambda$ for $\tau < \lambda$ and if there is a $(\lambda^{+},\lambda,\mu,\nu)$-family, then $\nu = \mu$ and $cf\mu = cf\lambda$.

PROOF. (i) is just a restatement of Theorem 1.3. We prove (ii).

Assume $2^{\tau} \leq \lambda$ for $\tau < \lambda$ and that $\mathcal{F}$ is a $(\lambda^{+},\lambda,\mu,\nu)$-family. If $\nu \neq \mu$, then there is $\mu'$ such that $\nu \leq \mu' \leq \mu$ and $cf\mu' \neq cf\lambda$. For each $A \in \mathcal{F}$, choose $A'\subseteq A$ such that $|A'| = \mu'$. Then $\mathcal{F}'= \{A': A \in \mathcal{F}\}$ is a

$(\lambda^+,\lambda,\mu',\nu)$-family. Since $cf\lambda \neq cf\mu'$, it follows by Corollary 1.6 that there is a cardinal $\tau < \lambda$ and an almost disjoint $(\lambda^+,\tau,\mu')$-family. But this contradicts the hypothesis $2^\tau \leq \lambda$. Therefore $\nu = \mu$. Likewise, if $cf\mu \neq cf\lambda$, it follows from Corollary 1.6 that $\lambda$ is not the least cardinal $\tau$ for which there is an almost disjoint $(\lambda^+,\tau,\mu)$-family and this is again a contradiction. □

The next theorem due of BAUMGARTNER [1976] shows that the smallest cardinal $\kappa$ for which there is no almost disjoint $(\kappa,\lambda,\lambda)$-family is regular.

THEOREM 1.10. <u>If $\kappa$ is a singular cardinal and if for every $\kappa' < \kappa$ there is a $(\kappa',\lambda,\lambda,\lambda)$-family, then there is also a $(\kappa,\lambda,\lambda,\lambda)$-family.</u>

<u>PROOF.</u> Choose cardinals $\kappa_\alpha < \kappa$ $(\alpha < cf\kappa)$ such that $\kappa = \bigcup\{\kappa_\alpha: \alpha < cf\kappa\}$. By hypothesis there is a $(cf\kappa,\lambda,\lambda,\lambda)$-family $\{X_\alpha: \alpha < cf\kappa\}$, and for each $\alpha < cf\kappa$ there is a $(\kappa_\alpha,\lambda,\lambda,\lambda)$-family, $\mathcal{F}_\alpha$, of subsets of $X_\alpha$. Then $\mathcal{F} = \bigcup\{\mathcal{F}_\alpha: \alpha < cf\kappa\}$ is a $(\kappa,\lambda,\lambda,\lambda)$-family. □

By Corollary 1.2 there is an almost disjoint $(2^\omega,\omega,\omega)$-family, and the first natural question one might ask is whether there is an almost disjoint $(2^{\omega_1},\omega_1,-)$-family. Of course, if $2^\omega = 2^{\omega_1}$, this is obvious. We also noted that (by Corollary 1.2) there is such a family if CH holds. The next theorem (BAUMGARTNER [1976]) is a strengthening of this.

THEOREM 1.11. <u>If</u> $2^\omega < 2^{\omega_1}$ <u>and if</u> $2^\omega < \aleph_{\omega_1}$, <u>then there is a</u> $(2^{\omega_1},\omega_1,\omega_1,\omega_1)$-<u>family</u>.

<u>PROOF.</u> By Theorem 1.10 it will be enough to show that there is a $(\kappa,\omega_1,\omega_1,\omega_1)$-family for every regular cardinal $\kappa$ such that $2^\omega < \kappa \leq 2^{\omega_1}$. Since $2^\omega < 2^{\omega_1}$, it follows from Corollary 1.2 that there is a $(2^{\omega_1},2^\omega,\omega_1,\omega_1)$-family. Let $\kappa$ be a fixed regular cardinal, $2^\omega < \kappa \leq 2^{\omega_1}$, and let $\lambda$ be the least cardinal for which there is a $(\kappa,\lambda,\omega_1,\omega_1)$-family. Then $\lambda \leq 2^\omega$ and $cf\lambda = \kappa$ or $cf\lambda = \omega_1$ by Corollary 1.6. Therefore, since $\lambda < \aleph_{\omega_1}$ and $\lambda < \kappa$, it follows that $\lambda = \omega_1$. □

In contrast to Theorem 1.11, BAUMGARTNER [1976] has shown that it is consistent that: $2^\omega = \aleph_{\omega_1}$, $2^{\omega_1} = \aleph_{\omega_1+1}$ <u>and there is no almost disjoint</u> $(2^{\omega_1},\omega_1,-)$-<u>family</u>. By Theorem 1.3, there is an almost disjoint

$(\omega_2,\omega_1,\omega_1)$-family, but it is consistent that: $2^{\omega} = 2^{\omega_1} = \omega_3$ and there is no almost disjoint $(\omega_3,\omega_1,\omega_1)$-family.

We conclude this section by mentioning one further result due to SHELAH [1971] and BAUMGARTNER [1976]. We have already seen (Theorem 1.9(ii)) that, if GCH holds, then there is no $(\kappa,\lambda,\lambda,\nu)$-family if $\kappa > \lambda > \nu$. The next theorem shows that, even without GCH, the existence of such a family implies that the cardinals are rather special.

THEOREM 1.12. Let $\nu \geq \omega$ and let $\lambda$ be the smallest cardinal such that there is a $(\kappa,\lambda,\lambda,\nu)$-family with $\kappa > \lambda + 2^{\nu}$. Then $cf\lambda = \omega$ and $\omega_{\lambda} = \lambda$. Also there is $\rho < \lambda$ such that $\rho^{\nu} \geq \kappa$.

PROOF. Suppose for a contradiction that $cf\lambda > \omega$. Let $\mathcal{F}$ be a $(\kappa,\lambda,\lambda,\nu)$-family, where $\kappa > \lambda + 2^{\nu}$. We can assume that $\kappa$ is regular. For each $A \in \mathcal{F}$ there is $\alpha_A$ such that $\nu \leq \alpha_A < \lambda$ and $|\alpha_A \cap A| = |\alpha_A|$ (put $\xi_{\emptyset} = \nu$, $\xi_{n+1} = \sup\{\xi\colon |\xi \cap A| < |\xi_n|\}$, then $\alpha_A = \sup \xi_n$ will do). There is $\mathcal{F}' \subseteq \mathcal{F}$ such that $|\mathcal{F}'| = \kappa$ and $\alpha_A = \alpha$ for all $A \in \mathcal{F}'$. Then $\{\alpha \cap A\colon A \in \mathcal{F}'\}$ is a $(\kappa,\alpha,|\alpha|,\nu)$-family, contrary to the definition of $\lambda$. Therefore, $cf\lambda = \omega$.

Now let $\mu$ be any regular cardinal, $\nu < \mu < \lambda$. For $A \in \mathcal{F}$, let $A(\mu)$ denote the first $\mu$ elements of $A$ (in the ordering of $\lambda$). Then $\{A(\mu)\colon A \in \mathcal{F}\}$ is a $(\kappa,\lambda,\mu,\nu)$-family. Let $\rho$ be minimal such that there is a $(\kappa,\rho,\mu,\nu)$-family. Then $\rho \leq \lambda$. Also $cf\rho = \mu$ by Theorem 1.4, and so $\rho < \lambda$. By the minimality of $\lambda$, it follows that $\mu < \rho$ and hence $\omega_{\mu} \leq \rho < \lambda$. Therefore, $\omega_{\alpha} < \lambda$ for every $\alpha < \lambda$ and so $\lambda = \omega_{\lambda}$. Also, by Corollary 1.8, $\kappa \leq \rho^{\nu}$. □

However, BAUMGARTNER (1976) proved that both the existence, and the non-existence of an $(\omega_2,\omega_1,\omega_1,\omega)$-family is consistent with $2^{\omega} \geq \omega_2$.

## 2 MARTIN'S AXIOM AND ALMOST DISOINT SETS

We have seen (Corollary 1.2) that there is an almost disjoint $(2^{\omega},\omega,\omega)$-family. If CH is false, a natural question to ask is whether there is a maximal almost disjoint $(\omega_1,\omega,\omega)$-family. This turns out to be independent of the usual axioms of set theory. To discuss this, and some related questions, we give a brief introduction to Martin's axiom.

Let $\mathcal{P} = \langle P,\leq\rangle$ be a partially ordered set. Two elements $p,q \in P$ are compatible if there is an element $r \in P$ such that $p \leq r$ and $q \leq r$. We say that P is directed if every pair of elements in P are compatible. The partial order $\mathcal{P}$ is CCC if P contains no uncountable set of incompatible elements (CCC historically stands for "countable chain condition", but it is a condition on the size of incompatible sets). A subset $D \subseteq P$ is cofinal in P if $(\forall p \in P)(\exists d \in D) p \leq d$. If $\mathcal{F}$ is a family of cofinal subsets of P, we call a set $G \subseteq P$ $\mathcal{F}$-generic if G is directed and $F \cap G \neq \emptyset$ for every $F \in \mathcal{F}$. MA($\kappa$) is the assertion: whenever $\mathcal{P}$ is a CCC partially ordered set and $\mathcal{F}$ is a family of $\leq \kappa$ cofinal subsets of $\mathcal{P}$, then there is an $\mathcal{F}$-generic set. It is easily seen that MA($\omega$) is true (G can be chosen to be an $\omega$-chain). However, MA($2^\omega$) is false (see e.g. KUNEN [1980],p.54), so MA($\kappa$) implies $\kappa < 2^\omega$. Martin's axiom is the statement

$$\text{MA: MA}(\kappa) \text{ is true for every } \kappa,\ \omega \leq \kappa < 2^\omega.$$

Although this axiom does not have the same intuitive appeal as the other axioms of ZF, it does arise quite naturally in the context of forcing. However, it is not necessary to be familiar with the theory of forcing in order to apply the axiom, which turns out to be a powerful combinatorial principle with applications to algebra, analysis and point set topology. In fact it is equivalent to the topological assertion: a compact Hausdorff space not containing uncountably many disjoint open sets is not the union of fewer than $2^\omega$ closed nowhere dense sets (see e.g. KUNEN [1980],p. 65). Clearly CH implies MA, but the usefulness of MA is that it can frequently be employed as an alternative to CH, which is important in view of the fundamental result of SOLOVAY & TENNENBAUM [1971]: It is consistent that MA holds and $2^\omega > \omega_1$.

Many outstanding problems can be settled with the aid of MA. For example MA($\omega_1$) easily implies the Suslin hypothesis (SOLOVAY & TENNENBAUM [1971], see also KUNEN [1980], p74). However, here we shall be more concerned with the interplay between MA and almost disjoint families , in particular with the way MA is used (see Theorem 2.4 below) to encode subsets of $\kappa$ ($< 2^\omega$) by subsets of $\omega$ (a technique due to Solovay described in MARTIN & SOLOVAY [1970] and JENSEN & SOLOVAY [1970]). This will be an immediate consequence of the following important result.

THEOREM 2.1. Assume MA($\kappa$). If $\mathcal{A}, \mathcal{B} \subseteq \mathcal{P}(\omega)$ are such that $|\mathcal{A} \cup \mathcal{B}| \leq \kappa$ and $|B \setminus \cup\mathcal{A}'| = \omega$ whenever $B \in \mathcal{B}$ and $\mathcal{A}'$ is a finite subset of $\mathcal{A}$, then there is a set $C \subseteq \omega$ such that $C \cap A$ is finite and $C \cap B$ is infinite for all $A \in \mathcal{A}$ and $B \in \mathcal{B}$.

PROOF. Let $\mathcal{P} = \langle P, \leq \rangle$ be the partial order on the set P of all pairs $(s,\mathcal{F})$, where $s \in [\omega]^{<\omega}$, $\mathcal{F} \in [\mathcal{A}]^{<\omega}$ and $(s_1,\mathcal{F}_1) \leq (s_2,\mathcal{F}_2) \Leftrightarrow s_1 \subseteq s_2$, $\mathcal{F}_1 \subseteq \mathcal{F}_2$ and $\bigcup\mathcal{F}_1 \cap s_2 \subseteq s_1$. Note that two elements $(s,\mathcal{F})$, $(s',\mathcal{F}')$ are compatible in $\mathcal{P}$ if and only if

$$\bigcup\mathcal{F} \cap s' \subseteq s \quad \text{and} \quad \bigcup\mathcal{F}' \cap s \subseteq s',$$

in which case $(s \cup s', \mathcal{F} \cup \mathcal{F}')$ is a common extension. $\mathcal{P}$ is clearly CCC since in any uncountable subset of $\mathcal{P}$ there are two elements $(s,\mathcal{F})$ and $(s,\mathcal{F}')$ which have the same first component, and which are therefore compatible.

For $A \in \mathcal{A}$, let $D_A = \{(s,\mathcal{F}): A \in \mathcal{F}\}$, and for $B \in \mathcal{B}$ and $n \in \omega$, let $E_{B,n} = \{(s,\mathcal{F}): s \cap B \setminus n \neq \emptyset\}$. It is easy to check that each $D_A$ and each $E_{B,n}$ is cofinal in $\mathcal{P}$ and so by $MA(\kappa)$ there is a directed set, G, which has a non-empty intersection with all these sets. The theorem holds with $C = \bigcup\{s: (s,\mathcal{F}) \in G$ for some $\mathcal{F}\}$. For, if $A \in \mathcal{A}$, there is $(s,\mathcal{F}) \in G \cap D_A$ and, since any $(s',\mathcal{F}') \in G$ is compatible with $(s,\mathcal{F})$, it follows that $A \cap s' \subseteq \bigcup\mathcal{F} \cap s' \subseteq s$, and so $A \cap C$ is finite. If $B \in \mathcal{B}$, then, for any $n \in \omega$, there is $(s,\mathcal{F}) \in G \cap E_{B,n}$ and so $C \cap B \setminus n \supseteq s \cap B \setminus n \neq \emptyset$, and therefore $C \cap B$ is infinite. □

Theorem 2.1 has many applications. For example, it immediately answers the question raised at the beginning of this section that is, it is consistent that there is no maximal almost disjoint $(\omega_1,\omega,\omega)$-family. More generally,

THEOREM 2.2. <u>MA implies that there is no maximal almost disjoint $(\kappa,\omega,\omega)$-family for $\omega \leq \kappa < 2^\omega$.</u>

PROOF. Given an almost disjoint $(\kappa,\omega,\omega)$-family ,$\mathcal{A}$, by Theorem 2.1 (with $\mathcal{B} = \{\omega\}$), there is an infinite set $C \subseteq \omega$ almost disjoint from each $A \in \mathcal{A}$. □

We should mention that, in contrast to Theorem 2.2, it is also consistent that: <u>for each cardinal $\kappa$ $(\omega_1 \leq \kappa \leq 2^\omega)$ there is a maximal almost disjoint $(\kappa,\omega,\omega)$-family</u> - see the remark following 3.6 below.

We shall use the following consequence of Theorem 2.1. For subsets A,B of $\omega$, write $A \blacktriangleright\!\blacktriangleright B$ if $A \setminus B$ is infinite and $B \setminus A$ is finite. Obviously, there is no strictly decreasing sequence of subsets of $\omega$ of length $\omega_1$. However, ...

THEOREM 2.3. <u>MA implies that there is a $\blacktriangleright\!\blacktriangleright$-decreasing sequence of subsets of $\omega$ of length $2^\omega$.</u>

PROOF. Let $\alpha < 2^\omega$ and suppose we have already defined $A_\beta$ for $\beta < \alpha$ so that $A_\gamma \blacktriangleright\blacktriangleright A_\beta$ holds for $\gamma < \beta < \alpha$. Then by Theorem 2.1, with $\mathcal{A} = \{\omega \setminus A_\beta: \beta<\alpha\}$ and $\mathcal{B} = \{A_\beta: \beta<\alpha\}$, there is a set $A_\alpha$ such that $A_\beta \blacktriangleright\blacktriangleright A_\alpha$ for all $\beta < \alpha$. □

One of the important uses of Theorem 2.1 is that it gives a way to code the subsets of a cardinal $\kappa$ ( $< 2^\omega$) by subsets of $\omega$. Let $\mathcal{F} = \{A_\alpha: \alpha<\kappa\}$ be a fixed almost disjoint $(\kappa,\omega,\omega)$-family. We call a set $C \subseteq \omega$ an <u>$\mathcal{F}$-code</u> for a subset $S \subseteq \kappa$ if, for every $\alpha \in \kappa$,

$$C \cap A_\alpha \text{ is infinite } \Leftrightarrow \alpha \in S.$$

THEOREM 2.4. <u>Every set $S \subseteq \kappa$ has an $\mathcal{F}$-code</u>.

PROOF. Apply Theorem 2.1 with $\mathcal{A} = \{A_\alpha: \alpha \notin S\}$ and $\mathcal{B} = \{A_\alpha: \alpha \in S\}$. □

Note that since the $\mathcal{F}$-code mapping $\mathcal{P}(\kappa) \to \mathcal{P}(\omega)$ is clearly 1-1, we obtain the following corollary.

COROLLARY 2.5. <u>MA implies that $2^\kappa = 2^\omega$ for $\omega \leq \kappa < 2^\omega$</u>.

We conclude this section by proving a surprising result of KUNEN [1969] which uses the above results about MA and almost disjoint sets. MANSFIELD [1967] answered a question of ULAM by showing that there are subsets of $\mathbb{R}\times\mathbb{R}$ which do not belong to the $\sigma$-algebra generated by CA rectangles (i.e. rectangles $X\times Y$ with $X$, $Y$ co-analytic sets of reals). This led Kunen to consider the question whether every subset of the plane is in the larger $\sigma$-algebra generated by all the rectangles $X\times Y$ ($X, Y \subseteq \mathbb{R}$). The answer depends upon which axioms of set theory one chooses.

THEOREM 2.6. <u>If MA holds then every set $T \subseteq \mathbb{R}\times\mathbb{R}$ is expressible in the form</u>

$$T = \bigcap_{n\in\omega} \bigcup_{m\geq n} X_m \times Y_m,$$

<u>where</u> $X_m, Y_m \subseteq \mathbb{R}$.

Note that it is a theorem of ZFC that every subset of $\omega_1 \times \omega_1$ can be expressed in the stated form (with $X_m, Y_m \subseteq \omega_1$) (KUNEN [1969]), and so CH implies Theorem 2.6. To prove Theorem 2.6 without assuming CH we use a special coding. By Theorem 2.3 there is a $\blacktriangleright\blacktriangleright$-decreasing sequence $(A_\alpha: \alpha < 2^\omega)$ of $2^\omega$ subsets of $\omega$. Put $B_\alpha = A_\alpha \setminus A_{\alpha+1}$. Then $\{B_\alpha: \alpha < 2^\omega\}$ is an almost disjoint $(2^\omega,\omega,\omega)$-family.

LEMMA 2.7. <u>Let</u> $S \subseteq 2^\omega$, $\alpha < 2^\omega$. <u>Then there is a set</u> $C \subseteq \omega$ <u>such that</u> (i) $C\cap B_\beta$ <u>is finite for all</u> $\beta > \alpha$ <u>and</u> (ii) <u>for</u> $\beta \leq \alpha$, $C \cap B_\beta$ <u>is infinite</u> $\Leftrightarrow$ $\beta \in S$.

PROOF. Apply Theorem 2.1 with $\mathcal{A} = \{B_\beta: \beta \le \alpha \text{ and } \beta \notin S\} \cup \{A_{\alpha+1}\}$ and $\mathcal{B} = \{B_\beta: \beta \le \alpha \text{ and } \beta \in S\}$. □

PROOF OF THEOREM 2.6. We identify $\mathbb{R}$ with $2^\omega$. First we show that any set $T \subseteq \{(\alpha,\beta): \beta \le \alpha < 2^\omega\}$ is expressible in the stated form. For each $\alpha < 2^\omega$, let $T_\alpha = \{\beta \le \alpha: (\alpha,\beta) \in T\}$. By Lemma 2.7 with $S = T_\alpha$, it follows that there is a set $C_\alpha \subseteq \omega$ such that $C_\alpha \cap B_\beta$ is infinite $\Leftrightarrow \beta \in T_\alpha$. Then

$$T = \bigcap_{n\in\omega} \bigcup_{m\ge n} \{(\alpha,\beta): m \in C_\alpha \cap B_\beta\} = \bigcap_{n\in\omega} \bigcup_{m\ge n} \{\alpha : m \in C_\alpha\} \times \{\beta : m \in B_\beta\}.$$

For $(\alpha,\beta) \in T \Leftrightarrow C_\alpha \cap B_\beta$ is infinite $\Leftrightarrow$ for every $n \in \omega$ there is $m \ge n$ such that $m \in C_\alpha \cap B_\beta$. By symmetry, any set $T' \subseteq \{(\alpha,\beta): \alpha \le \beta < 2^\omega\}$ can also be expressed in the same form. The theorem follows from this since

$$\bigcap_{n<\omega} \bigcup_{m\ge n} X_m \times Y_m = \bigcap_{n<\omega} \bigcup_{m\in e(n)} X_m \times Y_m \cup \bigcap_{n<\omega} \bigcup_{m\in o(n)} X_m \times Y_m,$$

where $e(n)$, $o(n)$ denote the even and odd integers $\ge n$. □

On the other hand, KUNEN [1969] also gives the following theorem.

THEOREM 2.8. <u>If there is a real-valued measurable cardinal $\kappa \le 2^\omega$, then there are planar sets not in the $\sigma$-algebra generated by rectangles $X \times Y$ with $X,Y \subseteq \mathbb{R}$.</u>

An uncountable cardinal $\kappa$ is <u>real-valued measurable</u> if there is a $\kappa$-additive real-valued measure, m, defined on $\mathcal{P}(\kappa)$ such that $m(\kappa) = 1$ and $m(S) = 0$ for $S \in [\kappa]^{<\kappa}$. It is not known if it is consistent that there is such a cardinal (if $V = L$, there is no such $\kappa$, SCOTT (1961)). Of course, by Theorems 2.6 and 2.8, <u>if MA holds then there is no real-valued measurable cardinal $\kappa \le 2^\omega$.</u>

<u>PROOF OF THEOREM 2.8</u>. Suppose there is a real-valued measurable cardinal $\kappa \le 2^\omega$, and that m is an appropriate measure on $\mathcal{P}(\kappa)$. We claim that the set $D = \{(\alpha,\beta): \beta < \alpha < \kappa\}$ is not in the $\sigma$-algebra generated by rectangles $X \times Y$ with $X,Y \subseteq \kappa$. Indeed, suppose for contradiction that D is in this $\sigma$-algebra. Then D is $m\times m$-measurable, where $m\times m$ is the product measure on $\kappa\times\kappa$. Let $D_\alpha = \{\beta: \beta < \alpha\}$, $D^\beta = \{\alpha: \beta < \alpha < \kappa\}$. Then $m(D_\alpha) = 0$ and $m(D^\beta) = 1$ and so, by Fubini's Theorem, we have the contradiction

$$0 = \int m(D_\alpha)dm(\alpha) = m\times m(D) = \int m(D^\beta)dm(\beta) = 1.$$

## 3 MAXIMAL ALMOST DISJOINT FAMILIES

An almost disjoint family of sets $\mathcal{F}$ is $\lambda$-MAD ($\lambda$-maximal almost disjoint) if $|\cup\mathcal{F}| = \lambda$, $|F| = \lambda$ $(\forall F \in \mathcal{F})$, and $\mathcal{F}$ is maximal in the sense that, for every set $X \in [\cup\mathcal{F}]^{\lambda}$ there is $F_X \in \mathcal{F}$ such that $|X \cap F_X| = \lambda$. Obviously, if $\delta < cf\lambda$ and $\mathcal{F}$ is a family of $\delta$ pairwise disjoint sets each of cardinality $\lambda$, then $\mathcal{F}$ is $\lambda$-MAD. On the other hand, there is no $\lambda$-MAD family of cardinality $cf\lambda$. For, if $\mathcal{F} = \{F_\alpha: \alpha < cf\lambda\}$ is an almost disjoint $(cf\lambda,\lambda,\lambda)$-family, then $A = \cup\{A_\alpha: \alpha < cf\lambda\}$ is almost disjoint from each member of $\mathcal{F}$ and has cardinality $\lambda$, where $A_\alpha \in [F_\alpha \setminus \cup\{F_\beta: \beta < \alpha\}]^{\lambda(\alpha)}$ and $\lambda(\alpha)$ $(\alpha < cf\lambda)$ is a sequence of cardinals cofinal in $\lambda$. Thus in particular, for regular $\lambda$, there is no $\lambda$-MAD family of size $\lambda$ and a natural question to ask is whether there is a $\lambda$-MAD family of cardinality $\lambda$ when $\lambda$ is singular. In this section we give a partial solution of this questiuon due to ERDÖS & HECHLER [1973], who attribute the question to W.W. COMFORT.

We write $\delta \in MAD(\lambda)$ if there is a $\lambda$-MAD family of cardinality $\delta$. The following lemma of HECHLER [1972] is similar to Theorem 1.10.

LEMMA 3.1. <u>Let $\lambda$, $\delta$ be infinite cardinals, $\delta$ singular. If $\delta$ is a limit of cardinals in $MAD(\lambda)$, then $\delta \in MAD(\lambda)$.</u>

PROOF. By the hypothesis, there are cardinals $\delta_\alpha \in MAD(\lambda)$ $(\alpha < cf\delta)$ which are cofinal in $\delta$. We can assume that $\delta_0 > cf\delta$. Let $\mathcal{F}_0$ be a $\lambda$-MAD family of size $\delta_0$ and let $g : \mathcal{F}_0 \to cf\delta$ be any onto mapping. For each set $F \in \mathcal{F}$, let $\mathcal{F}_F$ be a $\lambda$-MAD family of subsets of $F$ such that $\cup\mathcal{F}_F = F$ and $|\mathcal{F}_F| = \delta_{g(F)}$. Then $\mathcal{F} = \cup\{\mathcal{F}_F: F \in \mathcal{F}_0\}$ is a $\lambda$-MAD family of size $\delta$. □

ERDÖS & HECHLER gave two ways of constructing $\lambda$-MAD families. These are described in Theorems 3.2 & 3.6.

THEOREM 3.2 <u>Let $\lambda$, $\mu$ be cardinals, $\lambda > \mu$ and $\lambda$ singular. Then there is $\delta \in MAD(\lambda)$ such that $\mu \le \delta \le \mu^{cf\lambda}$.</u>

PROOF. Let $\lambda_\alpha$ $(\alpha < cf\lambda)$ be an increasing sequence of regular cardinals which is cofinal in $\lambda$, $\lambda_\alpha > \mu$. Let $S_{\alpha\beta}$ $(\alpha < cf\lambda,\ \beta < \mu)$ be pairwise disjoint sets, $|S_{\alpha\beta}| = \lambda_\alpha$. Let $\mathcal{F}$ be the set of all functions $f \in [(cf\lambda)\times\mu)]^{cf\lambda}$, and let $\mathcal{B}$ be a maximal almost disjoint subfamily of $\mathcal{F}$. Then $\mu \le |\mathcal{B}| \le \mu^{cf\lambda}$. Let $\mathcal{A} = \{X_f: f \in \mathcal{B}\}$, where $X_f = \cup\{S_{\alpha\beta}: \langle\alpha,\beta\rangle \in f\}$. We claim that $\mathcal{A}$ is $\lambda$-MAD. Clearly $|X_f| = \lambda$ $(f \in \mathcal{B})$. Also, if $X$ is any subset of $\cup\mathcal{A}$ $(\subseteq \cup\{S_{\alpha\beta}: \alpha < cf\lambda,\ \beta < \mu\})$ of cardinality $\lambda$, then, for each

$\alpha < \mathrm{cf}\lambda$, there are $\xi = \xi(\alpha) < \mathrm{cf}\lambda$ and $\eta = \eta(\alpha) < \mu$ such that $|X \cap S_{\xi\eta}| \geq \lambda_\alpha$. Let $g = \{\langle\xi(\alpha),\eta(\alpha)\rangle : \alpha < \mathrm{cf}\lambda\}$. By the maximality of $\mathcal{B}$, there is $f \in \mathcal{B}$ such that $|f \cap g| = \mathrm{cf}\lambda$, and hence $|X \cap X_f| = \lambda$. □

COROLLARY 3.3. <u>If $\lambda$ is singular and $\mu^{\mathrm{cf}\lambda} < \lambda$ for every $\mu < \lambda$, then $\lambda \in \mathrm{MAD}(\lambda)$.</u>

By Corollary 3.3, it follows that there is a $\lambda$-MAD family of size $\lambda$ if $\lambda$ is a singular strong limit cardinal; in particular this is the case if GCH holds. It is not known if it is consistent that there is a negative answer to COMFORT's question, i.e. whether it is consistent that there is a singular cardinal $\lambda \notin \mathrm{MAD}(\lambda)$.

If GCH holds, we have an exact description of $\mathrm{MAD}(\lambda)$ (for a stronger result see BALANDA [1983]).

COROLLARY 3.4. <u>Assume GCH. Then</u>

$$\mathrm{MAD}(\lambda) = \{\delta : 0 < \delta \leq \lambda^+, \delta \text{ a cardinal} \neq \mathrm{cf}\lambda\}.$$

PROOF. We already observed that $\mathrm{cf}\lambda \notin \mathrm{MAD}(\lambda)$ and $\delta \in \mathrm{MAD}(\lambda)$ for $\delta < \mathrm{cf}\lambda$. Also $\lambda^+ \in \mathrm{MAD}(\lambda)$ by Theorem 1.1. Suppose $\delta$ is a cardinal such that $\mathrm{cf}\lambda < \delta \leq \lambda$. If $\delta$ is regular, then $\delta \in \mathrm{MAD}(\lambda)$ by Theorem 3.2, and if $\delta$ is singular, $\delta \in \mathrm{MAD}(\lambda)$ by Lemma 3.1. □

Another immediate corollary of Theorem 3.2 is the following.

COROLLARY 3.5. <u>If</u> $2^\omega = \omega_n < \omega_\omega$, <u>then</u> $\{\omega_\alpha : n \leq \alpha \leq \omega\} \subseteq \mathrm{MAD}(\omega_\omega)$.

PROOF. $\omega_m^{\ \omega} = \omega_m$ for $n \leq m < \omega$. □

We have seen (Corollary 3.3) that $\lambda \in \mathrm{MAD}(\lambda)$ if $\lambda$ is a strong limit cardinal. The next result enables one to say something when $2^{\mathrm{cf}\lambda} > \lambda$.

THEOREM 3.6. <u>$\mathrm{MAD}(\mathrm{cf}\lambda) \subseteq \mathrm{MAD}(\lambda)$.</u>

PROOF. Let $S_\alpha$ $(\alpha < \mathrm{cf}\lambda)$ be pairwise disjoint subsets of $\lambda$ such that $|S_\alpha| < \lambda = \cup\{S_\alpha : \alpha < \mathrm{cf}\lambda\}$. If $\mathcal{F}$ is a $(\mathrm{cf}\lambda)$-MAD family of subsets of $\mathrm{cf}\lambda$, then $\mathcal{F}' = \{\cup\{S_\alpha : \alpha \in F\} : F \in \mathcal{F}\}$ is $\lambda$-MAD, and $|\mathcal{F}| = |\mathcal{F}'|$. □

By a standard forcing argument (ERDÖS & HECHLER [1973]) it follows from Theorem 3.6 that: <u>it is consistent with ZFC that $\mathrm{cf}\lambda < \lambda < 2^{\mathrm{cf}\lambda}$ and there is a $\lambda$-MAD family of size $\delta$ for every $\delta \neq \mathrm{cf}\lambda$ in the range $0 < \delta \leq 2^{\mathrm{cf}\lambda}$</u>. We do not know if there is a converse of Theorem 3.6. For example, if $\omega_1 \in \mathrm{MAD}(\omega_\omega)$, does it necessarily follow that $\omega_1 \in \mathrm{MAD}(\omega)$ (and hence that $\mathrm{MA}(\omega_1)$ is false)?

## 4 A THEOREM OF GALVIN ON ČECH FUNCTIONS

A Čech function on $\kappa$ is a map $f: \mathcal{P}(\kappa) \to \mathcal{P}(\kappa)$ such that

(i) $(\forall X \subseteq \kappa)\ f(X) \supseteq X$,

(ii) $(\forall X,Y \subseteq \kappa)\ f(X \cup Y) = f(X) \cup f(Y)$,

(iii) $(\forall Y \subseteq \kappa)(\exists X \subseteq \kappa)\ f(Y) = X$,

(iv) $f$ is not the identity map.

ČECH [1947] asked if there is such a function, and the question is still unanswered by the usual axioms of ZFC.

Note that, if $f$ is a Čech function, it is increasing ($f(X) \subseteq f(Y)$ for $X \subseteq Y$) and $f(X) = X$ for every finite set. In fact, as J. VINARECK (see PRICE [1982]) observed, for any infinite set $A$, there is an infinite subset $B \subseteq A$ such that $f(C) = C$ for every set $C \subseteq B$. If the Čech function $f$ is $\kappa$-additive, i.e. if $f(\bigcup\{A_\alpha: \alpha < \lambda\}) = \bigcup\{f(A_\alpha): \alpha < \lambda\}$ for $\lambda < \kappa$, then $f(X) = X$ for every subset $X$ of cardinality $|X| < \kappa$. Denote by $\check{C}(\kappa)$ the assertion : <u>there is a $\kappa$-additive Čech function on $\kappa$.</u>

GALVIN [1978 a] announced that $\check{C}(\kappa)$ is true if $\kappa$ is regular and $2^\kappa = \kappa^+$, or if $\kappa = \omega$ and MA holds. Independently, at about the same time but by a different method, PRICE [1979] proved that $\check{C}(\omega)$ is true if the axiom IF($\omega$) below is satisfied. Call a family $\mathcal{F} \subseteq [\kappa]^\kappa$ <u>independent</u> if

$$|F_1 \cap F_2 \cap \ldots \cap F_r \cap (\kappa \setminus F_{r+1}) \cap \ldots \cap (\kappa \setminus F_n)| = \kappa$$

whenever $F_1, F_2, \ldots, F_n$ are distinct members of $\mathcal{F}$. HAUSDORFF [1936], generalizing a result of FICHTENHOLZ & KANTOROVICZ [1935], proved that there is an independent family of subsets of $\kappa$ of size $2^\kappa$ (see also KUNEN [1980], Exercise A5). Price's axiom is IF($\kappa$): <u>There is no maximal independent family $\mathcal{F} \subseteq [\kappa]^\kappa$ of cardinality less than $2^\kappa$.</u> PRICE [1979,1982] showed that MA $\Rightarrow$ IF($\omega$) $\Rightarrow$ $\check{C}(\omega)$. In this section, as an application of almost disjoint sets, we shall give another sufficient condition for the existence of a Čech function proved by GALVIN [1978 b]. We are grateful to Fred Galvin for allowing this previously unpublished proof to be included here.

Call an almost disjoint family $\mathcal{F} \subseteq [\kappa]^\kappa$ <u>completely separable</u> if, for every subset $X \subseteq \bigcup\mathcal{F}$,

EITHER (i) $X \in I(\mathcal{F}) = \{Y \subseteq \kappa: (\exists\ \mathcal{F}' \in [\mathcal{F}]^{<\kappa})\ Y \subseteq \bigcup\mathcal{F}'\}$,

OR (ii) $X \in D(\mathcal{F}) = \{Y \subseteq \kappa: |\{F \in \mathcal{F}: F \subseteq Y\}| = 2^\kappa\}$.

Note that, for a regular cardinal $\kappa$, the sets $I(\mathcal{F})$ and $D(\mathcal{F})$ are disjoint. Consider the following axioms.

CS($\kappa$): $\exists$ a completely separable family $\mathcal{F} \subseteq [\kappa]^{\kappa}$ of cardinality $2^{\kappa}$.

H($\kappa$): if $\mathcal{F}$ is $\kappa$-MAD and $|\mathcal{F}| \geq \kappa$, then $|\mathcal{F}| = 2^{\kappa}$.

We have already seen (Theorem 2.4) that MA implies $H(\omega)$, and it can be shown by the methods of SHELAH [1978] that $H(\kappa)$ is consistent. GALVIN [1978 b] proved the following theorem.

THEOREM 4.1. For regular $\kappa$, $H(\kappa) \Rightarrow CS(\kappa) \Rightarrow \check{C}(\kappa)$.

We give the proof in two stages.

LEMMA 4.2. If $\kappa$ is regular then $H(\kappa) \Rightarrow CS(\kappa)$.

PROOF. Let $\{A_\alpha: \alpha < 2^{\kappa}\}$ be a 1-1 enumeration of $\mathcal{P}(\kappa)$. Let $\mathcal{T} \subseteq [\kappa]^{\kappa}$ be a family of $\kappa$ pairwise disjoint subsets of $\kappa$ such that $\bigcup\mathcal{T} = \kappa$. We shall define sets $F_\alpha$ $(\alpha < 2^{\kappa})$ by transfinite induction in the following way. Let $\alpha < 2^{\kappa}$ and suppose that $F_\beta \in [\kappa]^{\kappa}$ has been defined for $\beta < \alpha$ so that $\mathcal{F}_\alpha = \mathcal{T}\cup\{F_\beta: \beta<\alpha\}$ is almost disjoint. If $A_\alpha \in I(\mathcal{F}_\alpha)$, put $F_\alpha = T$, where T is some fixed member of $\mathcal{T}$. If $A_\alpha \notin I(\mathcal{F}_\alpha)$, then $|A_\alpha| = \kappa$ and we can choose a set $F_\alpha \in [A_\alpha]^{\kappa}$ which is almost disjoint from all the members of $\mathcal{F}_\alpha$. This is possible by the hypothesis $H(\kappa)$ since the almost disjoint family $\{F\cap A_\alpha: F \in \mathcal{F}_\alpha, |F\cap A_\alpha| = \kappa\}$ is not maximal. The family $\mathcal{F} = \mathcal{T}\cup\{F_\alpha: \alpha<2^{\kappa}\}$ constructed in this way is clearly almost disjoint. Also $\mathcal{F}$ has cardinality $2^{\kappa}$. For, if $|\mathcal{F}| < 2^{\kappa}$, then $\mathcal{F}$ is not maximal by the assumption $H(\kappa)$, and so there is some $A_\alpha \in [\kappa]^{\kappa}$ which is almost disjoint from all the members of $\mathcal{F}$. Since $\kappa$ is regular, $A_\alpha \notin I(\mathcal{F})$ and therefore, $F_\alpha \subseteq A_\alpha$, and this is a contradiction. It remains to verify that, if $X \in \mathcal{P}(\kappa) \setminus I(\mathcal{F})$, then $X \in D(\mathcal{F})$. Since X does not belong to the ideal $I(\mathcal{F})$, it follows that $|X| = \kappa$ and for any set $Y \subseteq X$, either Y or $X\setminus Y$ is not in the ideal. Therefore, by the construction there are $2^{\kappa}$ different ordinals $\alpha < \kappa$ such that $F_\alpha \subseteq A_\alpha \subseteq X$, and hence $X \in D(\mathcal{F})$. □

LEMMA 4.3. If $\kappa$ is regular then $CS(\kappa) \Rightarrow \check{C}(\kappa)$.

PROOF. Inductively define families $\mathcal{S}_n \subseteq [\kappa]^{\kappa}$ for $n \in \omega$ as follows. Put $\mathcal{S}_0 = \{\kappa\}$. Let $n \in \omega$ and suppose $\mathcal{S}_n$ has been defined. For each $Y \in \mathcal{S}_n$ let $\mathcal{F}_Y \subseteq [Y]^{\kappa}$ be a completely separable family of subsets of Y such that $\bigcup\mathcal{F}_Y = Y$ and $|\mathcal{F}_Y| = 2^{\kappa}$, and define $\mathcal{S}_{n+1} = \bigcup\{\mathcal{F}_Y: Y \in \mathcal{S}_n\}$. This defines $\mathcal{S}_n$

for all $n\in\omega$. Clearly, for $n > \emptyset$, $\mathcal{F}_n$ is a completely separable family of subsets of $\kappa$ of size $2^\kappa$. Put

$$\mathcal{T}_n = \{X \in D(\mathcal{F}_{n+1}): X \subseteq Y \in \mathcal{F}_n\}.$$

Then $\mathcal{F}_n \subseteq \mathcal{T}_n \subseteq D(\mathcal{F}_{n+1})$. By the definition of $D(\mathcal{F}_{n+1})$, there is a 1-1 map $\varphi_n: \mathcal{T}_n \to \mathcal{F}_{n+1}$ such that $\varphi_n(X) \subseteq X$. We will show that the function $f:\mathcal{P}(\kappa) \to \mathcal{P}(\kappa)$ defined by

$$f(X) = X\cup\bigcup\{A \subseteq \kappa: (\exists n\in\omega)\ A \in \mathcal{T}_n \text{ and } X\cap\varphi_n(A) \in D(\mathcal{F}_{n+2})\},$$

is a $\kappa$-additive Čech function.

Clearly $f(X) \supseteq X$. To see that $f$ is not the identity map consider the set $Z = \varphi_0(\kappa)$ (which is defined since $\kappa \in \mathcal{T}_0$). Since $Z \in \mathcal{F}_1$, $Z \neq \kappa$. Also, since $Z = Z\cap\varphi_0(\kappa) \in \mathcal{F}_1 \subseteq D(\mathcal{F}_2)$, it follows that $f(Z) = \kappa \neq Z$. It is also easy to see that $f$ is $\kappa$-additive. For, if $X = \bigcup\{X_\alpha: \alpha < \lambda\} \subseteq \kappa$ and $\lambda < \kappa$, then $f(X) = \bigcup\{f(X_\alpha): \alpha < \lambda\}$ since, by the regularity of $\kappa$, $X\cap\varphi_n(A) \in D(\mathcal{F}_{n+2})$ holds if and only if $X_\alpha\cap\varphi_n(A) \in D(\mathcal{F}_{n+2})$ holds for some $\alpha < \kappa$. It remains to show that $f$ is a surjective map.

Let $Y \subseteq \kappa$. We have to show that $f(X) = Y$ for some $X \subseteq Y$.

For $n \in \omega$ define

$$\mathcal{F}_n = \{F \in \mathcal{F}_n: Y\cap F \in D(\mathcal{F}_{n+1}) \text{ and } (\forall i<n)(F \subseteq G \in \mathcal{F}_i \to Y\cap G \notin D(\mathcal{F}_{i+1}))\},$$

and put $\mathcal{F} = \bigcup\{\mathcal{F}_n: n \in \omega\}$. Note that, (***): <u>if</u> $F \in \mathcal{F}_n$ <u>and</u> $Y\cap F \in D(\mathcal{F}_{n+1})$, <u>then there are sets</u> $F_j \in \mathcal{F}_j$ $(j\leq n)$ <u>such that</u> $\kappa = F_0 \supseteq F_1 \supseteq \dots \supseteq F_n = F$ <u>and there is a unique index</u> $k \leq n$ <u>such that</u> $F \subseteq F_k \in \mathcal{F}_k$.

If $F \in \mathcal{F}_n$, then $Y\cap F \in \mathcal{T}_n$ and therefore $\varphi_n(Y\cap F)$ is defined. Put

$$X_0 = \bigcup\{\varphi_n(Y\cap F): F \in \mathcal{F}_n,\ n\in\omega\},$$
$$Y_0 = Y \setminus \bigcup\mathcal{F}.$$

Then $X_0$, $Y_0$ are disjoint subsets of $Y$. We will show that

(1) $$f(X_0\cup Y_0) = Y.$$

We first prove that

(2) $$f(Y_0) = Y_0.$$

To see this observe that, if $X$ is any subset of $\kappa$ such that $f(X) \neq X$, then there are $n \in \omega$ and $A \in \mathcal{T}_n$ such that $X\cap\varphi_n(A) \in D(\mathcal{F}_{n+2})$, and so $X \in D(\mathcal{F}_{n+2})$. Therefore, there is a least integer $m$ such that $X \in D(\mathcal{F}_m)$. Then $m \geq 1$ since $D(\mathcal{F}_0) = \emptyset$, and so $X \in I(\mathcal{F}_{m-1}) \setminus I(\mathcal{F}_m)$. Hence there is $E \in \mathcal{F}_{m-1}$ such that $X\cap E \notin I(\mathcal{F}_m)$, and therefore $X\cap E \in D(\mathcal{F}_m)$. It follows from (***) that there are $k < m$ and $F \in \mathcal{F}_k$ such that $E \subseteq F$. Thus $E \subseteq \bigcup\mathcal{F}$. Since $|X\cap E| = \kappa$ and $Y_0\cap E = \emptyset$ it follows that $X \neq Y_0$. This proves (2).

If $n \in \omega$ and $F \in \mathcal{F}_n$, then $Y \cap F \in \mathcal{T}_n$ and $X_o \cap \varphi_n(Y \cap F) = \varphi_n(Y \cap F) \in \mathfrak{F}_{n+1} \subseteq D(\mathfrak{F}_{n+2})$. Therefore, $Y \cap F \subseteq f(X_o)$. It follows that

$$\cup\{Y \cap F : F \in \mathcal{F}\} = Y \setminus Y_o \subseteq f(X_o).$$

From this and (2) and the fact that f is additive, it follows that

$$f(X_o \cup Y_o) = f(X_o) \cup Y_o \supseteq Y.$$

Thus, in order to prove (1) we need only to show that $f(X_o) \subseteq Y$. In other words, by the definition of $f(X_o)$, it will be enough to show that, if

(3) $$A \in \mathcal{T}_n \text{ and } X_o \cap \varphi_n(A) \in D(\mathfrak{F}_{n+2})$$

holds, then

(4) $$A \subseteq Y.$$

Suppose that (3) holds. Put $B = \varphi_n(A)$. Then $B \in \mathfrak{F}_{n+1}$ and $X_o \cap B \in D(\mathfrak{F}_{n+2})$. Therefore, $Y \cap B \in D(\mathfrak{F}_{n+2})$ and so by (***) there are $k \leq n+1$ and $F_k \in \mathcal{F}_k$ such that $B \subseteq F_k$. Since $F_k \in \mathcal{F}_k$ it follows that $Y \cap F_k \in D(\mathfrak{F}_{k+1})$ and $Y \cap F_i \notin D(\mathfrak{F}_{i+1})$ for $i < k$, where $F_i \in \mathfrak{F}_i$ and

$$\kappa = F_o \supseteq F_1 \supseteq \ldots \supseteq F_k.$$

Hence, $Y \cap F_i \in I(\mathfrak{F}_{i+1})$ for $i < k$, and so there are $\lambda_i < \kappa$ and distinct sets $H_{i\xi} \in \mathfrak{F}_{i+1}$ $(i < k,\ \xi < \lambda_i)$ such that

$$Y \cap F_i \subseteq \bigcup_{\xi < \lambda_i} H_{i\xi}.$$

Since $|Y \cap F_{i+1}| \geq |Y \cap F_k| = \kappa$ for $i < k$, and since $\mathfrak{F}_{i+1}$ is an almost disjoint family, it follows that $F_{i+1} = H_{i\xi}$ for some $\xi < \lambda_i$, and we may assume $F_{i+1} = H_{i0}$. We claim that

(5) $$\cup\{E \in \mathcal{F} : E \neq F_k\} \subseteq \cup\{H_{i\xi} : i<k,\ 0<\xi<\lambda_i\}.$$

To see this, consider any set $E \in \mathcal{F}$ different from $F_k$. Since $F_k \in \mathfrak{F}_k$ and $E \in \mathfrak{F}_r$ for some $r \in \omega$, it follows that there is a largest integer $j \leq k$ such that E and F are both subsets of $F_j$; in fact, $j \neq k$ by the definition of $\mathcal{F}$. Now $|Y \cap E| = \kappa$ (since $E \in \mathcal{F}$) and so $|Y \cap F_j| = \kappa$ and therefore, by the regularity of $\kappa$, $|Y \cap H_{j\xi}| = \kappa$ for some $\xi < \lambda_j$. Moreover, by the definition of j, E is not a subset of $F_{j+1} = H_{j0}$, and so $\xi \neq 0$. This proves (5).

From (5), the definition of $X_o$, and the fact that $\varphi_n(E) \subseteq E$ $(E \in \mathcal{T}_n)$, it follows that

$$X_o \setminus \varphi_k(Y \cap F_k) \subseteq \cup\{E \in \mathcal{F} : E \neq F_k\} \subseteq \cup\{H_{i\xi} : i<k,\ 0<\xi<\lambda_i\}.$$

Since $B \subseteq F_k \subseteq F_{i+1} = H_{i0}$ and $|H_{i0} \cap H_{i\xi}| < \kappa$ for $\xi \neq 0$, it follows that $|B \cap (X_o \setminus \varphi_k(Y \cap F_k))| < \kappa$. Therefore, $B \cap (X_o \setminus \varphi_k(Y \cap F_k))$ belongs to the ideal $I(\mathfrak{F}_{n+2})$. Since $B \cap X_o \in D(\mathfrak{F}_{n+2})$, it follows that $B \cap \varphi_k(Y \cap F_k)$, and hence $\varphi_k(Y \cap F_k)$, also belong to $D(\mathfrak{F}_{n+2})$. Thus $\varphi_k(Y \cap F_k) \in D(\mathfrak{F}_{n+2}) \cap \mathfrak{F}_{k+1}$. Since

$k \leq n+1$ and $D(\mathcal{F}_{n+2}) \cap \mathcal{F}_{n+2} = \emptyset$, it follows that $k \leq n$.

Suppose that $k = n$. Then $B \cap \varphi_n(Y \cap F_n) \in D(\mathcal{F}_{n+1})$ and so has cardinality $\kappa$. Therefore, since $B = \varphi_n(A)$ and $\varphi_n(Y \cap F_n)$ both belong to the almost disjoint family $\mathcal{F}_{n+1}$, and since $\varphi_n$ is 1-1, it follows that $A = Y \cap F_n \subseteq Y$ and (4) holds.

Finally, suppose that $k < n$. Since $A \in \mathcal{F}_n$, there is $Z \in \mathcal{F}_n$ such that $A \subseteq Z$, and there is $Z_1 \in \mathcal{F}_{k+1}$ such that $Z \subseteq Z_1$. Thus $B = \varphi_n(A) \subseteq A \subseteq Z_1$. Therefore, $Z_1 \cap \varphi_k(Y \cap F_k) \supseteq B \cap \varphi_k(Y \cap F_k) \in D(\mathcal{F}_{n+2})$ and so $|Z_1 \cap \varphi_k(Y \cap F_k)| = \kappa$. Since $Z_1$ and $\varphi_k(Y \cap F_k)$ are both members of the almost disjoint family $\mathcal{F}_{k+1}$, they must be identical. Therefore, $A \subseteq Z_1 = \varphi_k(Y \cap F_k) \subseteq Y \cap F_k \subseteq Y$, and again (4) holds. This completes the proof. □

## 5 CONCLUDING REMARKS

In this paper we have given the basic properties of almost disjoint families, which are well known to set-theorists, and we have also given two examples of their use from point-set topology. There are many other applications and problems that we should have mentioned, but instead we must refer the interested reader to other sources. For example, an interesting problem would be to characterize those families of sets, $\mathcal{F}$, which have almost disjoint refinements (i.e. a map $f: \mathcal{F} \to \mathcal{P}(\bigcup\mathcal{F})$ such that $f(F) \subseteq F$ and $\{f(F): F \in \mathcal{F}\}$ is almost disjoint). For progress on this problem we refer the reader to the paper by BALCAR & VOJTAS [1980] who showed that every uniform ultrafilter on $\omega$ has this property. The survey by JECH & PRIKRY [1979] shows the connection between almost disjoint functions (or transversals) and uniform ultrafilters and also shows how almost disjoint functions lead to an elementary proof of SILVER's [1974] theorem on singular cardinals. Chapter §12 of COMFORT & NEGREPONTIS [1974] gives other connections between almost disjoint families, Boolean algebras, ultrafilters and topology.

## References


Balander, K.P. (1983). Maximally almost disjoint families of representing sets, Maths.Proc. Cam.Phil.Soc., 93, 1-7.

Balcar, B & Vojtas, P (1980). Almost disjoint refinement of families of subsets of N. Proc. Amer. Maths. Soc., 79, 465-470.

Baumgartner, J.E. (1976). Almost-disjoint sets, the dense set problem and the partition calculus. Ann. Math.Logic, 10, 401-439.

Čech, E. (1947). Probleme, Fund. Math., 34, 332.

Comfort, W.W. & Negrepontis, S. (1974). The theory of ultrafilters, Springer-Verlag, Berlin.

Erdős, P. & Hechler, S.H. (1975). On Maximal almost disjoint families over singular cardinals. In Infinite and Finite Sets, Vol. I, ed. A. Hajnal et al. pp 597-604 (Colloq. Keszthely, 1973, Colloq. Math. Janos Bolyai Vol. 10), (North-Holland, Amsterdam).

Fichtenholz, G. & Kantorovitch, L. (1935). Sur les opérations linéairs dans l'espace des fonctions bornées. Studia Math. 5, 69-98.

Galvin, F. (1978a). On a problem of Čech. Notices Amer. Math. Soc., 25, A-604 and A-720.

Galvin, F. (1978b). Private communication.

Hausdorff, F. (1936). Über zwei Sätze von G. Fichtenholz und L. Kantorovitch. Studia Math 6, 18-19.

Hechler, S.H. (1972). Short complete nested sequences in $\beta N - N$ and small maximal almost disjoint families. General Topology and Appl.2, 139-149.

Jech, T. & Prikry, K. (1979). Ideals over uncountable sets: Applications of almost disjoint functions and generic ultrapowers. Amer. Math. Soc. Memoirs Vol. 18, No. 214.

Jensen, R.B. & Solovay, R.M. (1970). Some applications of almost disjoint sets. In Mathematical Logic and Foundations of Set Theory, ed. Y. Bar-Hillel, pp. 143-167. Amer. Math. Soc., Providence, Rhode Island.

Kunen, K. (1968). Inaccessibility properties of cardinals. Stanford University, Ph.D. Thesis.

Kunen, K. (1980). Set Theory, An Introduction to Independence Proofs. North Holland, Amsterdam.

Mansfield, R. (1967). The solution of one of Ulam's problems concerning analytic rectangles. Lecture notes for 1967 U.C.L.A. Summer Institue on Axiomatic Set Theory.

Martin, D. & Solovay, R.M. (1970). Internal Cohen extensions. Ann. Math Logic, 2, 143-178.

Mitchell, W. (1972). Aronszajn trees and the independence of the transfer property. Ann Math. Logic, 5, 21-46.

Price, R.A. (1979). CH (and less) solves a set theoretic problem of Čech. University of Wisconsin, Ph.D. Thesis.

Price, R.A. (1982). On a problem of Čech. General Topology and Appl., 14, 319-329.

Sierpinski, W. (1928). Sur une décomposition d'ensembles. Monatsch. Math. Phys. 35, 239-242.

Scott, D.S. (1961). Measurable cardinals and constructible sets. Bull. Acad. Polon. Sci. 9, 521-524.

Shelah, S. (1971). The number of non-isomorphic models of an unstable first-order theory. Israel J. Math., 9, 473-487.

Shelah, S. (1978). A weak generalization of MA to higher cardinals. Israel J. Math., 30, 297-306.

Silver, J.H. (1974). On the singular cardinals problem. Proc. Int. Congr. Math. Vancouver, 265-268.

Solovay, R.M. & Tennenbaum, S. (1971) Iterated Cohen extensions and Suslin's problem, Ann. of Math. 94, 201-245.

Tarski, A. (1928) Sur la décomposition des ensembles en sous-ensembles presque disjoints, Fund. Math., 12, 188-205.

Tarski, A. (1929). Sur la décomposition des ensembles en sous-ensembles presque disjoints, Fund. Math., 14, 205-215.

# Random Graphs, Strongly Regular Graphs and Pseudo-random Graphs

Andrew Thomason†
Department of Mathematics
University of Exeter
North Park Road
Exeter EX4 4QE
England

## 1. Introduction.

Despite the title this article is not an attempt to be all things to all men. For although random graphs and strongly regular graphs are often thought of as opposite ends of a spectrum, the one being chaotic and disordered, the other being structured and symmetric, there are certain occasions on which they might bear similarities. These occasions are broadly of two kinds:

(i) there are extremal graph theory problems where both random graphs and strongly regular graphs provide the extremal configurations, or, at least, the best known approximations, and

(ii) there are instances where graphs with certain desired properties can be shown to exist by random methods, but where the best constructed examples are strongly regular, or close to strongly regular.

(Of course, these two instances overlap somewhat.) On these occasions both types of graph may be thought of as being covered by the umbrella term 'pseudo-random graph'. The purpose of this paper is to try and make precise this similarity and to offer the elements of a common treatment for both types of graph.

**Caveat.** Let it be said at once that we can prove nothing about random graphs which cannot be proved better by standard methods, and our methods are of little effect in sparse graphs anyway. Likewise we tackle only a restricted class of strongly regular graphs, namely those with parameters $(k, \lambda, \mu)$ where $\lambda \approx \mu$ (or equivalently those whose two non-trivial eigenvalues are opposite in sign and approximately equal in magnitude). However for this class it is possible to obtain some new results.

The aim of a common treatment for the different types of graph in case (i) above would be to attempt a characterisation of the extremal graphs for the given problem, and so perhaps help to compute the extremal function. This approach was successful in the proof of Theorem 2.3 below and in establishing new upper bounds for ramsey numbers; it also produced unexpected results about another problem

† *Current address*: Department of Pure Mathematics and Mathematical Statistics, 16, Mill Lane, Cambridge CB2 1SB, England.

(discussed in section 6). The aim in case (ii) would be to provide a way to check whether a given construction is 'pseudo-random', and so has the desired property. Some results of this kind are discussed in section 5.

At present this discussion may appear a bit vague, but in section 3 we will give a precise definition of a class of graphs (*jumbled* graphs) which will be our class of pseudo-random graphs, and show that both random graphs and certain types of strongly regular graphs fall into this class. In section 4 we develop some properties of jumbled graphs and in section 6 look at some of the consequences. To begin with, though, we look at some of the instances of the problems mentioned at the outset.

For matters concerning random graphs we refer to the recent treatise of Bollobás [**12**]. The material in chapter 13 of that work is of particular relevance. The notation $\mathcal{G}(n,p)$ will refer to the probability space of labelled graphs on $n$ vertices where the edges are chosen independently and at random with probability $p$. We shall use the phrase 'a random graph has property $P$', where $P$ is a graph property, to mean that a graph in $\mathcal{G}(n,p)$ has $P$ almost surely, that is, $\Pr(G \in \mathcal{G}(n,p) \text{ has } P) \to 1$ as $n \to \infty$.

## 2. Some examples.

Here we give some examples of occasions when random graphs and strongly regular graphs appear similar.

RAMSEY THEORY.

The best illustration is provided by the first ever result on random graphs. Let $r(K_t)$ be the ramsey number of $K_t$, that is, the smallest value of $n$ such that every colouring of the edges of the complete graphs $K_n$ of order $n$ with two colours yields a monochromatic $K_t$.

**Theorem 2.1.** (Erdős [**28**], 1947) $\quad r(K_t) > 2^{t/2}$.

This lower bound is obtained because there is only a very small probability of finding a $K_t$ in a random graph of order $2^{t/2}$ or its complement. Remarkably, there is no known constructive lower bound which is provably of exponential order, the best being $r(K_t) > \exp((1+o(1))(\log t)^2/4\log\log t)$ by Frankl and Wilson [**35**]. When it comes to exact values, though, strongly regular graphs appear. The only exact values known are $r(K_3) = 6$ and $r(K_4) = 18$, the extremal colourings being provided by the Paley graphs $Q_5$ and $Q_{17}$. The *Paley graph* $Q_n$ of order $n$ has vertex set the finite field $\mathbf{F}_n$, $n \equiv 1 \pmod 4$, with $xy \in E(Q_n)$ if $x-y$ is a square in $\mathbf{F}_n$. Elementary character sums show that $Q_n$ is a conference graph; that is, it is $(n-1)/2$-regular, each edge is in $(n-5)/4$ triangles, and that the same holds for the complement. A conference graph is strongly regular with parameters $((n-1)/2, (n-5)/4, (n-1)/4)$.

Conference graphs resemble random graphs with edge probability $1/2$ in that each vertex has degree around $n/2$ and each pair of vertices has around $n/4$ common neighbours. This suggests that lower bounds for ramsey numbers are provided by random-like graphs. (Indeed, by considering the extremal graphs for the function $r(K_t)$ as being pseudo-random it is possible to obtain a new upper bound for $r(K_t)$, as described in section 6.) More generally it would suggest that if $n$ is large, a colouring of $K_n$ with few monochromatic $K_t$'s is random-like, leading to the following conjecture. Let $k_t(G)$ denote the number of complete subgraphs in $G$ of order $t$, and let $\overline{G}$ denote the complement of $G$. Let

$$c_t(n) = \min \{ \ k_t(G) + k_t(\overline{G}) \ ; \ |G| = n \ \} \binom{n}{t}^{-1}$$

where $|G|$ is the order of $G$. (We use the notation of [**11**].) So $c_t(n)$ is the minimum proportion of monochromatic $K_t$'s in a colouring of $K_n$. Then $r(K_t) = \min\{n; c_t(n) > 0\}$. It is easily shown that $c_t(n)$ increases with $n$ so $c_t = lim_{n \to \infty} c_t(n)$ exists.

**Conjecture.** (Erdős [**29**], 1962). $\quad c_t = 2^{1-\binom{t}{2}}$.

Note that $2^{1-\binom{t}{2}}$ is the proportion of monochromatic $K_t$'s in a random colouring. In fact it will be shown later (Theorem 4.8) that a conference graph yields the same proportion of monochromatic $K_t$'s. This suggests that the extremal graphs for the function $c_t(n)$ are 'pseudo-random', and so might be those characterised in section 3. Indeed the corresponding conjecture involving complete bipartite subgraphs of bipartite graphs has been completely proved by Erdős and Moon [**30**]. (The corresponding conjecture is that the minimum proportion of monochromatic subgraphs is the same as the average proportion.) Surprisingly, it turns out Erdős' conjecture is *false*; counterexamples can be constructed by modifying certain pseudo-random graphs. We will return to this in section 6.

THE PROBLEM OF ZARANKIEWICZ.

This extremal problem (or a typical case of it) is that of computing the function $Z(n,t)$, the greatest number of edges in a graph of order $n$ which contains no complete bipartite subgraph $K_{t,t}$. For a general discussion see Bollobás [**11**]. Upper bounds on $Z(n,t)$ are due to Kövári, Sós and Turán [**49**] and Znám [**72**].

**Theorem 2.2.** $\quad \frac{1}{2}n^{2-2/(t+1)} < Z(n,t) < \frac{1}{2}(t-1)^{1/t}n^{2-1/t} + \frac{1}{4}(t-1)n.$

The general lower bound is obtained from random graphs with the appropriate number of edges. For $t = 2$ a better lower bound is obtained by considering graphs constructed from finite geometries. The graph constructed with vertex set

PG(2, $q$) by joining $a{:}b{:}c$ to $\alpha{:}\beta{:}\gamma$ if $a\alpha + b\beta + c\gamma = 0$ (Erdős and Rényi [**31**]), shows $Z(n,2) = \frac{1}{2}n^{3/2}(1+o(1))$ if $n$ is large. This graph is not strongly regular, but the degrees differ by at most one and the number of common neighbours of a pair of vertices varies by at most one. In a similar vein Brown [**22**] constructed from AG(3, $q$) bipartite graphs with n vertices and order $n^{5/3}$ edges, containing no $K_{3,3}$. The vertices of this graph consist of two copies of AG(3, $q$), with $(x_1, x_2, x_3)$ joined to $(y_1, y_2, y_3)$ if $\sum_{i=1}^{3}(x_i - y_i)^2 = -s$, where $s$ is some fixed quadratic non-residue. The graph is $(q^2 - q)$-regular and no pair of vertices has more than $q+1$ common neighbours. At least to this extent the extremal graphs for the Zarankiewicz problem are pseudo-random.

SUBCONTRACTIONS.

The fundamental extremal problem involving subcontractions is to determine the greatest number of edges in a graph $G$ of given order which does not contract to a complete graph of order $t$; we write $G \not\succ K_t$. It is not hard to show that the function

$$b(t) \;=\; \liminf_{n\to\infty} \{\, b \;;\; |G| = n \text{ and } e(G) \ge b|G| \text{ implies } G \succ K_t \,\}$$

exists, and is at most $2^{t-3}$. Mader [**51**] proved $t-2 \le b(t) \le 8(t-2)\lfloor \log_2(t-2) \rfloor$ for $t \ge 4$. Only recently it was noticed that random graphs show $b(t) \ge \frac{1}{4}t\sqrt{\log_2 t}(1 + o(1))$ (de la Vega [**68**], Kostochka [**48**], Thomason [**61**]), and the latter two references contain proofs that $t\sqrt{\log t}$ is the correct order for $b(t)$. The next result is from [**61**].

**Theorem 2.3.** $0{\cdot}265t\sqrt{\log_2 t}(1+o(1)) \le b(t) \le 2{\cdot}68t\sqrt{\log_2 t}(1+o(1))$.

It might be expected that the extremal graphs for this problem are pseudo-random, and it was by examining subcontractions in pseudo-random graphs that the simple proof in [**61**] was discovered.

THE DIAMETER.

The function $n(D,\Delta)$, the greatest number of vertices in a graph of maximum degree $\Delta$ and diameter $D$ seems very hard to determine. It is easily shown that $n(D,\Delta) \le (\Delta(\Delta-1)^D - 2)/(\Delta - 2)$. From below the de Bruijn graphs show $n(D,\Delta) \ge \lfloor \Delta/2 \rfloor^D$. The vertices of the de Bruijn graph $dB(k,d)$ are the $k^d$ vectors with $d$ coordinates which are integers between 1 and $k$. Two vectors are adjacent in the graph if the first $k-1$ coordinates of one agree with the last $k-1$ coordinates of the other. Most of the best results about $n(D,\Delta)$ are quite recent. For good surveys covering different aspects of this area see Bermond and Bollobás [**7**], Bermond, Bond, Paoli and Peyrat [**8**] and Bollobás [**12**]. From our point of view we note that random regular graphs provide the best known bounds for many instances of $n(D,\Delta)$. This is natural in that the paths from a vertex in an extremal graph must spread out as much as possible, and that is a property of random regular graphs. The extremal graphs are pseudo-random but this is a case where the methods of section 3 do not apply easily, because the edge density is too low, and so we can make no worthwhile contribution.

EXPANDER GRAPHS.

A bipartite graph with vertex classes $X$ and $Y$ each of order $n$ is $(n, a, b)$-*expanding* if every subset $A \subset X$ of order $a$ has at least $b$ neighbours in $Y$. It is an $(n, k, \beta)$-*expander* if it is $(n, x, x(1 + \beta(1 - x/n)))$-expanding for all $x \leq n/2$. Much interest has been shown lately in graphs with few edges which are good expanders. They are used for instance in the construction of concentrators and superconcentrators (Margulis [**52**], Valiant [**67**], Chung [**25**]), and the expanding properties of graphs are used implicitly or explicitly in the construction of parallel sorting algorithms (Häggkvist and Hell [**43**], Bollobás and Thomason [**19**], Ajtai, Komlós and Szemerédi [**1**], Alon [**3**]). Random graphs provide the best upper bounds known for the size of expanders (see Chung [**25**]). Many constructions have been found, some such as that of Margulis [**52**] requiring deep techniques for proof. The construction of concentrators requires expanders whose sizes are linear in the number of vertices, and such graphs are again too sparse for our techniques to be amenable. But sometimes a dense expander needs to be constructed (say for sorting applications) and here we can make some contribution. We will return to this in section 5.

## 3. Jumbled graphs.

In a random graph in $\mathcal{G}(n, p)$, each induced subgraph $H$ satisfies $e(H) \approx p\binom{|H|}{2}$, where $e(H)$ is the number of edges of $H$. The least we might require of a graph which mimics a random graph of edge probability $p$ is that the same property holds. To this end, we call the quantity $\left|e(H) - p\binom{|H|}{2}\right|$ the *error* of the subgraph $H$, and we will define a pseudo-random graph to be one in which no induced subgraph has large error. It is the choice of error term which determines the usefulness of the definition; the one we choose was described in [**62**].

Definition. *Let $p, \alpha$ be real numbers with $0 < p < 1 \leq \alpha$. A graph $G$ is said to be $(p, \alpha)$-*jumbled *if every induced subgraph $H$ satisfies*

$$\left|e(H) - p\binom{|H|}{2}\right| \leq \alpha|H|.$$

The reason for choosing this form of error term is first of all that it depends only on $H$, so implying

(a) if $G$ is $(p, \alpha)$-jumbled then $\overline{G}$ is $(1 - p, \alpha)$-jumbled, and

(b) if $G$ is $(p,\alpha)$-jumbled and $G'$ is an induced subgraph of $G$ then $G'$ is $(p,\alpha)$-jumbled.

The form of error term also enables us to prove Theorems 3.1 and 3.2 below, which we will discuss later.

What is the significance of the parameter $\alpha$? Of course, every graph of order $n$ is $(p,n/2)$-jumbled but if a graph is known to be $(p,o(n))$-jumbled quite a lot can be said about its properties, as we shall see. A conference graph of order $n$ is $(p,\sqrt{n})$-jumbled by Theorem 3.1 below; in fact a theorem of Erdős and Spencer [**32**] (or more precisely a theorem proved by a method similar to a proof in [**32**]) shows that if $G$ is $(p,\alpha)$-jumbled then $\alpha \neq o(\sqrt{pn})$. Observe that the definition permits us to say nothing useful about subgraphs $H$ of $G$ if $p|H| = O(\alpha)$. Since we often need information about the neighbourhood of a vertex, which will usually have order around $pn$, we mostly need $\alpha = o(p^2n)$. Because of this inequality and the relation $\alpha \neq o(\sqrt{pn})$ our methods will be most easily applied to dense graphs, that is with $p \geq n^{-1/3}$. From time to time we shall use the term 'jumbled graph' by itself, without specifying the values of $p$ and $\alpha$, to mean a $(p,\alpha)$-jumbled graph where $\alpha$ is suitably small (say $\alpha = o(p^2n)$). The word 'jumbled' is intended to convey the fact that the edges are evenly spread through the graph.

We aim to show that jumbled graphs mimic in many ways the *large scale* properties of random graphs with edge probability $p$. But before describing the properties of jumbled graphs, let us see some examples of them. There are essentially two ways to test whether a graph is $(p,\alpha)$-jumbled for some small $\alpha$. First of all, each subgraph $H$ might be tested against the definition. In this way it can be shown that a random graph is $(p,2\sqrt{pn})$-jumbled; this is fortunate, for otherwise our class of pseudo-random graphs would not contain random graphs themselves. However we are not always put to so much trouble.

**Theorem 3.1.** *Let $G$ be a graph of order $n$, with minimum degree $pn$. If no pair of vertices has more than $p^2n + l$ common neighbours, $G$ is $(p,\sqrt{(p+l)n})$-jumbled.*

This is a remarkable (though simply proved) result, since it shows that the condition on vertex degrees and common neighbours, which we noticed in earlier examples, is actually sufficient to make a graph behave like a random graph. This theorem is very useful for showing that specific constructions, such as Paley graphs and others in the following list, are jumbled. The need to examine all induced subgraphs is removed, and all that remains is a simple degree check. (The proof of the theorem is indeed simple, depending only on the Cauchy-Schwartz inequality or second moment method, and so is probably implicit in several earlier works by other authors. In fact the conditions in the theorem imply a somewhat stronger conclusion, namely that the error of an induced subgraph $H$ is at most $(1+\sqrt{pn}+\sqrt{l|H|})|H|$. Under certain circumstances this would yield information about $H$ when the conclusion of Theorem 3.1 was too weak, for instance if $l \approx \sqrt{n}$ and $|H| \approx n^{5/8}$. Of course we could modify our definition of a jumbled graph to take account of this extra strength, for instance by calling a graph $(p, \alpha, \beta)$-jumbled if the error of any induced subgraph $H$ is at most $(\alpha+\beta\sqrt{|H|})|H|$. Our reason for not doing so, apart from the extra complication that would arise with a fancier error term, is that no correspondingly stronger version of Theorem 3.2 below would be obtained. In any case no occasion has arisen so far where the full strength of the proof of Theorem 3.1 was needed. *N.b.* A gap in the proof of Theorem 3.1 given in [**62**] is filled in [**66**].)

SOME JUMBLED GRAPHS.

(a) A random graph in $\mathcal{G}(n,p)$ is $(p, 2\sqrt{pn})$-jumbled.

(b) A random graph in $\mathcal{G}(n,p)$, with edges added to form a clique of order $\sqrt{pn}$, is $(p, 3\sqrt{pn})$-jumbled.

(c) The vertex disjoint union of a random graph in $\mathcal{G}(n,p)$ and a clique $K_{\sqrt{pn}}$ is $(p, 3\sqrt{pn})$-jumbled.

The above examples show that the small subgraphs of a jumbled graph may be far from random.

(d) The Paley graphs $Q_n$ are $(1/2, \sqrt{n})$-jumbled.

(e) Let $n = 2kr + 1$ be a prime power. The graph whose vertices are the elements of the finite field $\mathbf{F}_n$, with $x$ joined to $y$ if $x - y$ is a $k$'th power, is $(1/k, 2n^{3/4})$-jumbled. This follows from Theorem 3.1 and estimates of Weil [**70**] for character sums. This graph is not strongly regular unless $k = 2$, when it is the Paley graph. If we let $xy$ be an edge if $x - y$ is in one of $j$ specified cosets of the $k$'th powers we obtain a graph which is $(j/k, 2n^{3/4})$-jumbled.

Hence we have specific constructions which emulate graphs in $\mathcal{G}(n,p)$ for any fixed rational value of $p$.

(f) The previous construction works if we join $x$ to $y$ whenever $x + y$ is a $k$'th power. This graph is not strongly regular even if $k = 2$. The example is interesting, though, because an obvious generalisation enables us to construct pseudo-random hypergraphs. This subject is explored further by Haviland and Thomason [**44**].

(g) Let the vertices of a graph be the vectors of the space $\mathrm{AG}(2,q)$, and partition the set of $q+1$ lines in this space into two sets $P$ and $N$, with $|P| = k$. Join $\mathbf{x}$ to $\mathbf{y}$ if $\mathbf{x} - \mathbf{y}$ is parallel to a line in $P$. Then $G$ is strongly regular with parameters $(k(q-1), (k-1)(k-2)+q-2, k(k-2))$, as recorded by Hubaut [**45**] and Seidel [**57**], and is $(k/q, n^{3/4})$-jumbled.

(h) Let the vertices of the graph $T_k^+$ be the $n = 2^{2k}$ vectors in AG$(2k, 2)$, with $\mathbf{x}$ joined to $\mathbf{y}$ if $q^+(\mathbf{x} - \mathbf{y}) \neq 0$, where

$$q^+((x_1, x_2, \ldots, x_{2k})) = x_1x_2 + x_3x_4 + \cdots + x_{2k-1}x_{2k}.$$

The graph $T_k^-$ is defined similarly by using $q^-$, where $q^-(\mathbf{x}) = x_1 + x_2 + q^+(\mathbf{x})$. The graphs $T_k^\pm$ are strongly regular with parameters $(2^{2k-1} \mp 2^{k-1}, 2^{2k-2} \mp 2^{k-1}, 2^{2k-2} \pm 2^{k-1})$ (see for example Thomason [63], Seidel [57] or Hubaut [45]), and so are $(1/2, n^{3/4})$-jumbled.

(i) Let the vertices of $G$ be the elements of PG$(k, q)$. Join the vertex $x_0{:}x_1{:}\ldots{:}x_k$ to the vertex $y_0{:}y_1{:}\ldots{:}y_k$ if $x_0y_0 + \cdots + x_ky_k = 0$. This graph is $(1/q, 2\sqrt{n/q})$-jumbled. When k=2 this graph is the Erdős-Rényi graph mentioned in section 2.

Note that $q \sim n^{1/k}$, so we have a way to model graphs in $\mathcal{G}(n, n^{-1/k})$.

(j) The previous example may be viewed, when $q = 2$, as the graph whose vertices are the non-empty subsets of a set of order $k + 1$, two vertices being adjacent if their intersection has even order. Let $G$ be the subgraph spanned by the subsets of even order. Then $G$ is $(1/2, 2\sqrt{n})$-jumbled, where $n = 2^k - 1$. In fact if k is even then $G$ is strongly regular, with parameters $((n-3)/2, (n-11)/4, (n-3)/4)$.

(k) In the previous example we could have looked at the subgraph spanned by the vertices of odd order. This is also $(1/2, 2\sqrt{n})$-jumbled.

(l) Let the vertices of the graph $B(n, t)$ be the elements of the field $\mathbf{F}_n$, where $n$ is prime, and let $t$ be an integer, $1 < t < n$. Join $x$ to $y$ if the fractional part of $(x - y)^2/n$ is at most $t/n$. A theorem of Bollobás quoted in section 5 implies this graph is $(t/n, 3n^{3/4}\log n)$-jumbled.

(m) Let $G$ be a graph of order $r$, and let $m \geq 1$ be an integer. Denote by $m{\circ}G$ the graph of order $mr$ obtained by taking $r$ disjoint sets of vertices $V_x$, $x \in G$, with $|V_x| = m$, and joining $v_x \in V_x$ to $v_y \in V_y$ if $x$ is joined to $y$ in $G$. Note that $m_1{\circ}(m_2{\circ}G) = (m_1m_2){\circ}G$. If $G$ is $(p, \alpha)$-jumbled then $m{\circ}G$ is $(p, m\alpha + m)$-jumbled.

(n) The graph $2{\bullet}G$ is formed from two disjoint copies $G_1$ and $G_2$ of $G$; if $x_1 \in G_1$ and $y_2 \in G_2$ then $x_1y_2 \in E(2{\bullet}G)$ if $x = y$ or $xy \notin E(G)$. For suitable choices of $G$ this graph provides good lower bounds for ramsey numbers, as Mathon [53] showed. However there is no profit in iterating the operation $2\bullet$, since $2{\bullet}(2{\bullet}G)$ is isomorphic to $2{\bullet}(2{\circ}G)$.

These are just a few graphs we have selected, either because they are well known, or because they will be used as examples later, or simply to illustrate how easy it is to find examples of jumbled graphs. There are many others. Some more are given in [62], and there are many more strongly regular graphs with $\lambda \approx \mu$, such as those listed in [**45**] or some new ones of Brouwer [**21**]. The point is that Theorem 3.1 gives a very easy way of checking whether a given graph is jumbled.

From the point of view of describing the extremal graphs for certain problems, the following theorem from [**62**] gives a property of graphs which are *not* jumbled.

**Theorem 3.2.** *Let $G$ be a graph of order $n$, let $\eta n$ be an integer between 2 and $n-2$, and let $\omega > 1$ be a real number. Suppose each induced subgraph $H$ of order $\eta n$ satisfies $\left|e(H) - p\binom{\eta n}{2}\right| \le \eta n \alpha$. Then $G$ is $(p, 7\sqrt{n\alpha/\eta}/(1-\eta))$-jumbled. Moreover $G$ contains an induced subgraph $G^*$ of order at least $\left(1 - \frac{880}{\eta(1-\eta)^2\omega}\right) n$ which is $(p, \omega\alpha)$-jumbled.*

The proof is by no means as straightforward as that of Theorem 3.1. To see how this theorem might be used, let $G_1, G_2, \ldots$ be a sequence of graphs with $|G_n| = n$, and let $\eta$ be a constant between zero and one. Let $\alpha(n)$ be any function of $n$ satisfying $\alpha(n) = o(n)$, and choose $\omega(n)$ so that $\omega(n)\alpha(n) = o(n)$ and $\omega(n) \to \infty$. Then the theorem shows that there is a constant $\delta = \delta(\eta)$ such that either $G_n$ contains an induced subgraph $G_n^*$ of order $(1 + o(1))n$ which is $(p, o(n))$-jumbled (which is often as good as $G_n$ itself being jumbled), or $G_n$ contains an induced subgraph $H$ of order $\lfloor \eta n \rfloor$ with $\left|e(H) - p\binom{|H|}{2}\right| > \delta n^2$. An example of the use of this theorem occurs in section 6.

## 4. Properties of jumbled graphs.

Here we give some properties of jumbled graphs which illustrate their claim to be pseudo-random. The first few are basic properties of vertex degrees.

**Theorem 4.1.** *Let $G$ be a $(p, \alpha)$-jumbled graph of order $n$, and let $0 < \epsilon < 1$. Then at least $(1-\epsilon)n$ of the vertex degrees of $G$ lie in the range $p(n-1) \pm 10\alpha\epsilon^{-1}$.*

**Theorem 4.2.** *Let $G$ be a $(p, \alpha)$-jumbled graph of order $n$, and let $0 < \epsilon < 1$. Let $H$ be an induced subgraph of $G$ of order $k$. Then at least $n - \epsilon k$ of the vertices of $G$ have between $pk - 21\alpha\epsilon^{-1}$ and $pk + 21\alpha\epsilon^{-1}$ neighbours in $H$.*

Under the degree conditions of Theorem 3.1 (by which phrase we shall speak of a graph, such as a conference graph, with minimum degree $pn$, in which no pair of vertices has more than $p^2 n + l$ common neighbours for some small $l$), it is possible to show that almost all sets of $k$ vertices have around $p^k n$ common neighbours. For general graphs these conditions don't apply, but we do have the following result. In this the number of vertices joined to every vertex in a set $U_1$ and to no vertex in a set $U_2$ is denoted $v(U_1, U_2)$.

**Theorem 4.3.** *Let $G$ be a $(p,\alpha)$-jumbled graph of order $n$, let $k,l \geq 0$ be integers and let $0 < \epsilon < 1$. Then $|v(U_1,U_2) - p^k q^l n| < 21(k+l)^2 \alpha \epsilon^{-1}$ for at least $(1-\epsilon)\binom{n}{k}\binom{n-k}{l}$ choices of sets $U_1$ and $U_2$ with $|U_1| = k$ and $|U_2| = l$. (Here $q = 1-p$.)*

These degree conditions and the definition of a $(p,\alpha)$-jumbled graph enable us to establish many analogues of well known random graph properties for jumbled graphs. We list just a few of the more obvious ones.

THE DIAMETER.

It is easily seen that a random graph in $\mathcal{G}(n,p)$ has diameter 2 if $p^2 n - 2\log n \to \infty$. For jumbled graphs we come fairly close.

**Theorem 4.4.** *Let $G$ be a $(p,\alpha)$-jumbled graph. Let $u,w \in G$ be vertices with degree at least $d$. If $pd > 4\alpha$ there is a $u$-$v$ path of length at most 3 in $G$. In particular, if $\delta(G) > 4\alpha p^{-1}$ then $G$ has diameter at most 3.*

THE CONNECTIVITY.

For a random graph in $\mathcal{G}(n,p)$ the vertex connectivity equals the minimum degree (see Bollobás and Thomason [**20**]).

**Theorem 4.5.** *Let $G$ be a $(p,\alpha)$-jumbled graph of order $n$. Then $\kappa(G) > \delta(G) - 4\alpha p^{-1} + 1$.*

For graphs satisfying the degree conditions of Theorem 3.1 this can be improved to $\kappa(G) = \delta(G)$.

HAMILTON CYCLES.

A random graph is hamiltonian if $pn - \log n - \log\log n \to \infty$ (this is not easy to prove; see [**12**]). The same is certainly not true for a jumbled graph if $p$ is small, even if we impose a minimum degree condition; consider example (c). But if $p$ is larger we can make progress.

**Theorem 4.6.** *Let $G$ be a $(p,\alpha)$-jumbled graph of order $n$, with minimum degree at least $pn$. If $(p-k/n)^2 n \geq 6(\alpha + 2k)$, where $k$ is a non-negative integer, then $G$ has a set of $k+1$ edge disjoint hamilton cycles.*

**Theorem 4.7.** *Let $G$ be a $(p,\alpha)$-jumbled graph of order $n$, with minimum degree $pn \geq m = \lceil 6\alpha p^{-1} \rceil$. Then $G$ has at least $\frac{1}{2}(pn)!/m!$ hamilton cycles.*

These theorems show for instance the existence of an exponentially large number of hamilton cycles, and a set of $(n/100)$ edge disjoint hamilton cycles, in the Paley graphs. This answers a question of Calkin [**24**]. Theorems 4.6 and 4.7 are proved using the Chvátal-Erdős theorem [**26**] to generate hamilton cycles, rather than applying the flipping method commonly used in random graph results. This method has also been applied to find a linear expected time hamilton cycle algorithm for graphs in $\mathcal{G}(n,p)$ if $p \geq 12n^{-1/3}$; see [**64**].

INDUCED SUBGRAPHS, CLIQUES AND THE CHROMATIC NUMBER.

We turn now to induced subgraphs, since one of our original aims was to investigate the apparently extremal graphs for Erdős' conjecture. Previous methods for estimating complete subgraphs have been rather *ad hoc*. Several authors (Blanchard [**9**], Bollobás and Thomason [**18**], Graham and Spencer [**39**]) have shown that $k_t(Q_n) = 2^{-\binom{t}{2}}\binom{n}{t}(1+O(n^{-1/2}))$ by using Weil's estimates [**70**] for character sums. For $t = 4$ the exact result, $k_4(Q_n) = n(n-1)((n-5)(n-17)+4(a^2-1))/1536$, where $n = a^2+b^2$ and $a$ is odd and coprime to $n$, was obtained by Evans, Pulham and Sheehan [**34**] and Thomason [**60**]. The only general result so far was due to Giraud [**37**] who shows that $k_4(G)+k_4(\overline{G}) = \frac{1}{32}\binom{n}{4}(1+O(n^{-1/2}))$ if $G$ is a conference graph. The following theorem extends this to a much larger class of graphs, and for all values of $t$.

**Theorem 4.8.** *Let $G$ be a $(p,\alpha)$-jumbled graph of order $n$, where $p \leq 1/2$. Let $F$ be a graph of order $r \geq 3$ with $m$ edges, and let $A$ be the order of its automorphism group. Suppose $\epsilon$ satisfies $0 < \epsilon < 1$ and $\epsilon^2 p^r n \geq 42\alpha r^2$. Then the number of induced subgraphs of $G$ isomorphic to $F$ lies between $(1-\epsilon)^r p^m q^{\binom{r}{2}-m} A^{-1} n^r$ and $(1+\epsilon)^r p^m q^{\binom{r}{2}-m} A^{-1} n^r$, where $q = 1-p$.*

Rosenfeld asked if a strongly regular graph could be found containing a given graph $F$ as an induced subgraph. A graph containing *every* graph of order $r$ as an induced subgraph was called *r-full* by Bollobás and Thomason [**18**], who showed that the Paley graphs of large order are $r$-full. Theorem 4.8 combined with Theorem 3.1 offers a great many more examples.

Theorem 4.8 shows that in a $(1/2, O(\sqrt{n}))$-jumbled graph the number of complete subgraphs is around $2^{-\binom{t}{2}}\binom{n}{t}$ for $t$ up to about $\frac{1}{2}\log_2 n$. It would be reasonable to hope to push this up to $\log_2 n$. But although this may be true for the Paley graphs, as we see at the end of section 6, it is in general untrue. In the graph of example (j) the number of $K_t$'s constructed by first choosing $k/2$ independent mutually orthogonal vectors and then choosing $r = t-k/2$ more in the space spanned by the first $k/2$ is of order $2^{-\binom{t}{2}+\binom{r}{2}}\binom{n}{t}$, as shown in [**62**]. So Theorem 4.8 cannot be improved in general.

If we ask for the clique number of a $(1/2, O(\sqrt{n}))$-jumbled graph, Theorem 4.8 shows it is at least $\frac{1}{2}\log_2 n$. No example is known where it is so small, but the clique number of example (k) is only $\log_2 n$. On the other hand the clique number may be as large as $\sqrt{n}$, as many of our examples show. Of course it will not be much larger since it follows directly from the definition that the clique number of a $(p, \alpha)$-jumbled graph is at most $1 + 2\alpha(1-p)^{-1}$. This contrasts with random graphs in $\mathcal{G}(n, 1/2)$, where the clique number is known always to within one and usually exactly (see Matula [**54**] or Bollobás and Erdős [**17**]). Estimates for the chromatic number are related to those of the clique number, so by a greedy algorithm we can colour a $(1/2, O(\sqrt{n}))$-jumbled graph with at most $2n/\log_2 n$ colours, the corresponding value for random graphs being $n/\log_2 n$ (Grimmett and McDiarmid [**40**]). However, there is a lower bound of $n/2\log_2 n$ for the chromatic number of a random graph, whereas the chromatic number of a $(1/2, O(\sqrt{n}))$-jumbled graph may be as low as $\sqrt{n}$. This is the case, for example, in the Paley graph $Q_n$ if $n$ is a perfect square.

Subcontractions and topological cliques.

The *contraction clique number* $ccl(G)$ of a graph $G$ is the largest value of $t$ for which $G \succ K_t$. The *topological clique number* $tcl(G)$ is the largest value of $t$ for which $G$ contains a subdivision of $K_t$. Of course, $ccl(G) \geq tcl(G)$. The values of $ccl(G)$ and $tcl(G)$ were investigated by Bollobás, Catlin and Erdős [**15**] and by Bollobás and Catlin [**14**] in relation to the conjectures of Hadwiger and Hajós. (The former conjectured $ccl(G) \geq \chi(G)$, which holds for almost every graph, the latter speculated that $tcl(G) \geq \chi(G)$, which fails for almost every graph.) They found $ccl(G) = n/\sqrt{\log_b n}(1 + o(1))$, where $b = 1/(1-p)$, and $tcl(G) = 2\sqrt{n/(1-p)}(1 + o(1))$ almost surely, for $G \in \mathcal{G}(n,p)$, with $p$ constant. Use of the theorems at the beginning of this section and the techniques of [**61**] gives the following.

**Theorem 4.9.** *Let $p, C$ be constants, let $G_n$ be a $(p, C\sqrt{n})$-jumbled graph of order $n$, and let $b = 1/(1-p)$. Then, as $n \to \infty$,*

$$ccl(G_n) \geq (1 + o(1))n/\sqrt{\log_b n}$$

*and*

$$2(1+C)(1-p)^{-1}\sqrt{n} \geq tcl(G_n) \geq (1 + o(1))\sqrt{pn}.$$

A more general result for $tcl(G)$ is available in [**62**]. We give no upper bound for $ccl(G)$. For consider the graph $G$ with vertex set $\{x_i, y_i; 1 \leq i \leq n/2\}$, where $x_i y_i \in E(G)$ and for each pair $1 \leq i < j \leq n/2$ the vertex $x_i$ is joined to exactly one of $x_j$ and $y_j$ chosen at random, and so is $y_i$. Then $G$ is $(1/2, 2\sqrt{n})$-jumbled, and indeed the graphs spanned by $\{x_1, \ldots, x_{n/2}\}$ and $\{y_1, \ldots, y_{n/2}\}$ are randomly chosen members of $\mathcal{G}(n/2, 1/2)$. But $ccl(G) \geq n/2$. In fact $ccl(G)$ cannot be much bigger than this since the clique number must be at least $ccl(G) - n/2$.

## 5. Other techniques.

The methods used to obtain results in sections 3 and 4 were elementary though complicated. Other techniques have been used on jumbled graphs, apart from those alluded to so far. Here we describe briefly one or two of them. For the Paley graphs, Bollobás used the method of Gauss sums to obtain the following.

**Theorem 5.1.** (Bollobás [**13**]) *The number of edges between a set of $k$ vertices of the Paley graph $Q_n$ and another disjoint set of $l$ vertices lies between $\frac{1}{2}kl - \frac{1}{2}\sqrt{kln}$ and $\frac{1}{2}kl + \frac{1}{2}\sqrt{kln}$.*

This improves on Theorem 4.3 for this particular graph. In particular if $n$ is large, we see $v(U_1, U_2) \neq 0$. This implies that every first order graph property is either possessed by almost every graph in $\mathcal{G}(n,p)$ and by $Q_n$ for all large $n$ or is possessed by almost no graph in $\mathcal{G}(n,p)$ nor by $Q_n$ for all large $n$ (see for instance [**12**]).

The theorem also shows that Theorem 3.1 cannot be significantly improved if we require say every three vertices to have around $p^3n$ common neighbours. For $Q_n$ contains complete subgraphs of order $\sqrt{n}$ if $n$ is a perfect square, and so is not $({}^1\!/_2, \alpha)$-jumbled for $\alpha < \sqrt{n}/2$. Moreover none of the other results of section 4 can be improved significantly for this graph.

Recall the graph $B(n,t)$ of example (l). This graph is clearly regular of degree $d = |\{x \in \mathbf{F}_n; x^2 \in \{1, \ldots, t\}\}|$. If $t \gg n^{1/4} \log n$ then $d/t \to 1$; in fact the Pólya-Vinogradov inequality (see Ayoub [**6**]) states $|d - t| < \sqrt{n} \log n$, and a deep improvement by Burgess [**23**] reduces the $\sqrt{n}$ to $At^{1-1/(r+1)}n^{1/4r}$, where $A$ is an absolute constant and $r$ is any positive integer. The following theorem of Bollobás gives bounds on the number of common neighbours a pair of vertices might have. The pleasing proof is based on that of Pólya.

**Theorem 5.2.** (Bollobás [**12**]) *No two vertices of the graph $B(n,t)$ have more than $t^2/n + \sqrt{n}\log^2 n$ common neighbours.*

Consequently the graph $B(n,t)$ can be shown to be $(p, 3n^{3/4}\log n)$-jumbled by Theorem 3.1. A universal analogue of this graph has been constructed by Bollobás and Erdős [**16**]. Denote by $R(n,\alpha,\delta)$ the graph with vertex set $\{1, \ldots, n\}$, in which $ij$ is an edge if the fractional part of $(i-j)^2\alpha$ is less than $\delta$; here we require $\alpha$ to be irrational and $0 < \delta < 1$ but $n$ need no longer be prime. Pinch used classical results of Hardy and Littlewood to prove a conjecture of Bollobás and Erdős about $R(n,\alpha,\delta)$, which shows that it is $(\delta, o(n))$-jumbled.

**Theorem 5.3.** (Pinch [**56**]) *For every irrational $\alpha$ there is a function $f_\alpha : \mathbf{N} \to \mathbf{N}$ such that $f_\alpha(n) = o(n)$, and such that no two vertices in any graph $R(n,\alpha,\delta)$ have more than $\delta^2 n + f_\alpha(n)$ common neighbours.*

Quite a few techniques have been developed for use on expander graphs. (Recall the definitions of section 2.) Margulis was the first to construct a family of linear expanders as follows. The vertex classes $X$ and $Y$ are both copies of $\mathbf{Z}_m \times \mathbf{Z}_m$. Join $(a,b)$ in $X$ to $(a,b)$, $(a+1,b)$, $(a,b+1)$, $(a,a+b)$ and $(-b,a)$ in $Y$, and call the resulting graph $M(m)$. Using fairly deep techniques from representation theory, Margulis proved an expansion property for these graphs.

**Theorem 5.4.** (Margulis [**52**]) *There is an absolute constant* $\beta_0$ *such that* $M(m)$ *is an* $(m^2, 5, \beta_0)$*-expander.*

Unfortunately no lower bound is provided for $\beta_0$. However variants of this construction have been shown to be good expanders by some authors, and Gabber and Galil [**36**] were able to give explicit values of $\beta$, using fourier analysis. These have been further refined; see for example Alon and Milman [**5**], Jimbo and Maruoka [**46**] and Alon, Galil and Milman [**4**] (among many others. The papers cited contain many references to the literature). A useful idea of Tanner [**59**], developed by Alon [**2**], gives a valuable sufficient condition for a graph to be a good expander. Alon was able to go considerably further, though, and show the necessity of the condition. A graph is a *strong* $(n,k,\beta)$*-expander* if it is $(n, x, x(1+\beta(1-x/n)))$-expanding for all $x \leq n$. Given a graph $G$ we define $\lambda(G)$ to be the second smallest eigenvalue of the matrix $D - A$, where $D$ is the diagonal matrix of vertex degrees and $A$ is the adjacency matrix. It is easily checked that if $G$ is regular the smallest eigenvalue is 0 and $\lambda(G) > 0$.

**Theorem 5.5.** (Alon [**2**]) *Let* $G$ *be a* $k$*-regular bipartite graph. If* $G$ *is a strong* $(n,k,\beta)$*-expander then* $\lambda \geq \beta^2/(1024 + 2\beta^2)$. *If* $\beta \leq (2d\lambda - \lambda^2)/d^2$ *then* $G$ *is a strong* $(n,k,\beta)$*-expander.*

This is a very useful result, especially since the best expanders are generated randomly and we can estimate the expansion properties very quickly by this method, though to compute the expansion properties exactly is known to be coNP-complete; see Blum, Karp, Vornberger, Papadimitriou and Yannakakis [**10**]. It is also possible to verify explicit constructions.

For the purposes of constructing superconcentrators very sparse (in fact linear) expanders are needed. But some applications, notably parallel sorting in few rounds, make use of dense expanders, and here the ideas of section 3 start to bear fruit. To sort $n$ objects which are ordered in some unknown way in just two rounds using $m$ parallel processors, $m$ pairs of objects, determined by some algorithm (which of course is just a graph of order $n$ and size $m$) are compared. All possible deductions are made by transitivity, and any pairs whose relative order is still hidden are then compared. The algorithm is successful if it leaves at most $m$ pairs to be compared in the second round. After Häggkvist and Hell [**43**] constructed an algorithm, Bollobás and Thomason [**19**] found the correct order for the minimal value of $m$ by showing that a random graph with $n^{3/2}\log n$ edges produces a succesful algorithm. Moreover a random graph with $n^{5/3}\log^{1/3} n$ edges produces a successful algorithm even if we allow only two step deductions between rounds (that is, we can deduce $a < b$ only if the first round yields $a < c$ and $c < b$ for some $c$). Once again, this is the correct order of magnitude. Known constructions for algorithms are less efficient. Alon [**3**] used eigenvalue methods to show that the bipartite graph whose vertex classes are the points and the hyperplanes of $\mathrm{AG}(d,q)$, adjacency in the graph reflecting incidence in the geometry, is $(n, x, n - n^{1+1/d}/x)$-expanding for all $x$. In the case $d = 4$ it is straightforward to take such an expander and construct a two round sorting algorithm with two step deductions using only $(22/3 + o(1))n^{3/4}$ edges, not far from optimal. The expanding properties of this graph can also be verified by the following analogue of Theorem 3.1 for bipartite graphs.

**Theorem 5.6.** *Let $G$ be a bipartite graph with vertex classes $X$ and $Y$, both of order $n$. Suppose each vertex in $X$ has degree at least $pn$ and that no two vertices of $X$ have more than $p^2n + l$ common neighbours. Then $G$ is $(n, x, n - (lx + pn)/p^2x)$-expanding for all $x$.*

This allows the construction of many dense expanders to be verified. For example, the graph due to Brown [**22**] described in section 2 is $(n, x, n - 6n^{1/2} - n^{4/3}/x)$-expanding for all $x$, and, as already mentioned, the graphs described in the previous paragraph are $(n, x, n - n^{1+1/d}/x)$-expanding for all $x$. The proof of the theorem can also be used to show the existence of a two round sorting algorithm with two step deductions using only $(3 + o(1))n^{3/4}$ edges. Further details are given in [**65**].

Theorem 5.6 was used by Dyer and Frieze [**27**] along with Theorem 3.1 to develop a polynomial expected time algorithm for a minimum cut. The algorithm seeks to find the minimum number of edges in a cut which partitions the vertices into two equal-sized subsets. The graphs are uniformly distributed among those of order 2n which have a 'small cut', that is, a cut with at most $(\frac{1}{2} - \epsilon)e(G)$ edges for some fixed $\epsilon > 0$. The theorems above are needed to prove that the cut, once found, is indeed minimum.

## 6. Ramsey theory.

In this final section we examine some consequences of the previous work for ramsey theory. Recall Erdős' conjecture from section 1, that $c_t = 2^{1-\binom{t}{2}}$. This conjecture is trivially true for $t = 2$, and for $t = 3$ follows at once from a result of Goodman [**38**]. However this case is much easier than $t \geq 4$ since, as Lorden

[50] showed, the number of monochromatic triangles depends only on the degree sequence. For $t \geq 4$ very little is known, the only general result being due to Giraud [**37**] who showed that $c_4 > \frac{1}{46}$.

It turns out this conjecture is false. Recall the definition of $m \circ G$ from example (m) of section 3.

**Lemma 6.1.** $k_t(m \circ G) + k_t(\overline{m \circ G}) = \sigma_t(G) 2^{1-\binom{t}{2}} \binom{n}{t} (1 + o(1))$,

where $$\sigma_t(G) = 2^{\binom{t}{2}-1} p^{-t} \Big\{ t! k_t(G) + \sum_{j=1}^{t} j! S(t,j) k_j(\overline{G}) \Big\}.$$

*Here $G$ has order $p$, $n = mp = |m \circ G|$, and the $o(1)$ term is with $p,t$ fixed and $m \to \infty$. $S(t,j)$ is a Stirling number of the second kind and represents the number of ways of partitioning $t$ labelled objects into $j$ non-empty parts.*

A consequence of this lemma is that any graph $G$ provides an upper bound for $c_t$, namely $c_t \leq 2^{1-\binom{t}{2}} \sigma_t(G)$. To find counterexamples to Erdős' conjecture we need graphs with $\sigma_t < 1$. These are by no means easy to find; certainly $\sigma_4(Q_n) > 1$ for the Paley graphs, as the calculation of $k_4(Q_n)$ cited in section 4 shows. But the graphs $T_k^-$ from example (h) do the trick. The graph spanned by the neighbours of the zero vector in $T_k^-$ is denoted $P_k^-$. The next results are from [**63**].

**Theorem 6.2.** $\sigma_4(P_4^-) < 0{\cdot}976$ *and* $\sigma_t(T_t^-) < 0{\cdot}936$ *for* $t \geq 5$.

**Corollary 6.3.** $c_t < 2^{1-\binom{t}{2}}$ *for* $t \geq 4$.

A modification of $2 \circ P_4^-$ can be used to show $c_4 < \frac{1}{33}$.

The failure of Erdős' conjecture is a considerable surprise, since it has always been thought that the best ramsey colourings for complete graphs were, roughly speaking, symmetric with respect to the two colours and such that the edges of any given colour were spread evenly through the graph. This is not true, in the following sense. Let us call a sequence $G_1, G_2, \ldots$ of graphs such that $|G_n| = n$ and $k_t(G_n) + k_t(\overline{G}_n) = c_t(n)\binom{n}{t}$ an *extremal sequence* for $K_t$. Clearly Corollary 6.3 means we cannot characterise the graphs of an extremal sequence as pseudo-random graphs, as we had once intended. But we can still uncover some properties of the extremal sequence. Theorem 4.8 shows that $G_n$ is not itself $(1/2, o(n))$-jumbled, nor indeed can it contain an induced subgraph of order $n + o(n)$ which is $(1/2, o(n))$-jumbled. The remarks following Theorem 3.2 now have the following consequence.

**Theorem 6.4.** *Let $G_1, G_2, \ldots$ be an extremal sequence for $K_t$, and let $\eta$ be a constant with $0 < \eta < 1$. Then there is a positive constant $\delta = \delta(\eta)$ such that $G_n$ contains an induced subgraph $H_n$ of order $\lfloor \eta n \rfloor$ with $\left| e(H) - \frac{1}{2}\binom{|H|}{2} \right| > \delta n^2$.*

In other words, the graphs of an extremal sequence contain large subgraphs with a significant bias towards one colour. Unfortunately this does not imply that the graphs themselves are biased toward one colour, since for example the complete bipartite graph $K_{n/2,n/2}$ contains large biased subgraphs.

These results have some bearing on the actual ramsey numbers for complete graphs. Székely defined the quantity

$$k(n) = \min\left\{ \sum_{t>0} k_t(G) + k_t(\overline{G})\,;\ |G| = n \right\}$$

and showed that estimates for $k(n)$ could be used to estimate ramsey numbers, as follows. Let $r(G,H)$ denote the ramsey number which is the smallest value of $n$ such that any colouring of the edges of $K_n$ with red and blue yields a red $G$ or a blue $H$. The number $r(G,G)$ is abbreviated $r(G)$. This coincides with our earlier definition of $r(K_t)$. The term $r(k,l)$ stands for $r(K_k, K_l)$. Székely's theorem gives lower bounds for ramsey numbers in terms of the function $k(n)$.

**Theorem 6.5.** (Székely, [**58**]) *Given $\epsilon > 0$ there is an $n_0$ such that if $n > n_0$ then*

*both* $n^{(1/2-\epsilon)s} \le k(n) \le \frac{1}{(k-2)!} r(k,2)r(k,3)\ldots r(k,k-3)r(k,k-2)r(k,k)^2$
*and* $n^{0\cdot 2275 \log n} \le k(n) \le n^{0\cdot 7214 \log n}$,

*where* $s = \max\{l; r(l,l) \le \sqrt{n}\}$.

Another function for which we can obtain a better bound is the ramsey number $r(C_4, K_n)$, where $C_4$ is a 4-cycle. It is easily checked that no two vertices of the Erdős-Rényi graph have two common neighbours. This means first that the graph contains no $C_4$, and secondly that we can apply Theorem 3.1 with $p = n^{-1/2} + O(n^{-1})$ and $l < n^{-1/2}$. Thus the graph is $(n^{-1/2}, 2n^{3/4})$-jumbled, and so contains no independent set of order $4n^{3/4}$. In fact, as remarked after Theorem 3.1, the proof yields somewhat more, namely that the independence number is at most $n^{3/4} + n^{1/2}$.

**Theorem 6.6.** $r(C_4, K_n) > (1 + o(1))n^{4/3}$.

This result was also obtained by Alon [**3**], using his eigenvalue method.

We conclude with some remarks about the ramsey numbers $r(K_n)$ themselves. The usual proof of the existence of ramsey numbers involves a local argument, that is, it is based on a discussion on vertex degrees. In this paper we have been looking at how a more global approach might be fruitful. In this context it is interesting to note some recent work of Gyárfás, Lehel, Schelp and Tuza [**41**], extended by Gyárfás, Lehel, Nešetřil, Rödl, Schelp and Tuza [**42**]. For a given graph $G$ they define $r_k^{\text{loc}}(G)$ to be the smallest value of $n$ for which any colouring of the edges of $K_n$ with any number of colours yields a monochromatic $G$ provided no more than $k$ colours appear at any vertex. (Of course, the first thing they have to do is show that $r_k^{\text{loc}}(G)$ exists.) We will then want to compare $r_2^{\text{loc}}(K_t)$ with $r(K_t)$. In general $r_2^{\text{loc}}(G)/r(G)$ can be arbitrarily large, though it is shown in [**41**] that if $G$ is connected then $r_2^{\text{loc}}(G)/r(G) < 3/2$. For complete graphs there is the following sharper result, in which the graph $K_m + \overline{K}_n$ consists of $n$ vertices each joined to every vertex of a $K_m$.

**Theorem 6.7.** (Gyárfás, Lehel, Schelp and Tuza [**41**]) $r_2^{\text{loc}}(K_m + \overline{K}_n) = r(K_m + \overline{K}_n)$ *if* $m \geq n - 1$.

In particular, $r(K_t) = r_2^{\text{loc}}(K_t)$. There are many other interesting comparisons and contrasts between local and global ramsey numbers contained in [**41**] and [**42**].

As for the ramsey number $r(K_m + \overline{K}_n)$, it was conjectured in [**60**] to be at most $2^m(m+n-2)+2$. This was backed up by a heuristic argument which *in vacuo* has some appeal but now appears hopeless, especially in view of the failure of Erdős' conjecture. The conjecture is true for $m = 1$ (trivially) and $m = 2$ (by Goodman's theorem, as is implicit in Walker [**69**]), but is indeed false for $m = 3$. It is shown in [**66**] that $P_k^-$ contains no $K_3 + \overline{K}_n$ if $n > 4^{k-2}$.

**Theorem 6.8.** $r(K_3 + \overline{K}_n) \geq 8n + 2\sqrt{n-1} - 7$ *if* $n = 4^k + 1$.

We mentioned earlier that the extremal colourings for $r(K_3)$ and $r(K_4)$ were provided by Paley graphs, and most of the hitherto best known bounds for small ramsey numbers were derived from these graphs (though recently Mathon [**53**] has improved these bounds considerably with other constructions, as mentioned in section 3). The actual clique number $cl(Q_n)$ in a Paley graph is unknown if $n$ is prime (though if $n$ is a perfect square the clique number is $\sqrt{n}$). Of course it is at least as large as the smallest non-residue, which value is sometimes at least $\epsilon \log n \log\log n$ for some $\epsilon > 0$ (Montgomery [**55**] assuming the Riemann hypothesis for all $L$-functions of real characters). In order to improve the lower bound for $r(K_t)$ given by Theorem 1.1 it would be necessary to show that $cl(Q_n) < 2\log_2 n$ infinitely often. As for lower bounds on $cl(Q_n)$, the results of [**9**],[**18**] and [**39**], or Theorem 4.9, show that $cl(Q_n) > \frac{1}{2}\log_2 n$, though this follows from the fact that $Q_n$ is self-complementary and $r(K_t) < 4^t$. Perhaps it would be possible to improve the methods of [**9**],[**18**] and [**39**] by replacing the estimates of Weil by the more recent estimates of Deligne (see Katz [**47**]), and so obtain $cl(Q_n) > \log_2 n$, but this would be a formidable undertaking. However we have the following general result from [**66**].

**Theorem 6.9.** *Let $G$ be a $(p,\alpha)$-jumbled graph of order $n$. Then $G$ contains $K_u + \overline{K}_w$, provided $w = \lceil p^u n - 2\alpha/(1-p)\rceil \geq 1$.*

If, in this theorem, $G$ is such that $w \geq r(t-u,t)$, then $G$ contains a monochromatic $K_t$. Use of the classical bound $r(k,l) < \binom{k+l-2}{k-1}$ due to Erdős and Szekeres [**33**] yields the next theorem, which in particular can be applied to $Q_n$.

**Theorem 6.10.** *Let $G_1$, $G_2$, ... be a sequence of graphs in which $G_n$ has order $n$ and is $(p, O(\sqrt{n}))$-jumbled. Then $k_t(G_n) + k_t(\overline{G}_n) > 0$ if $t < \frac{5}{8}\log_2 n(1+o(1))$.*

This theorem means that if the extremal colourings for the ramsey number of $K_t$ are pseudo-random then $r(K_t) < (3{\cdot}05)^t$. In general no upper bound of the form $r(K_t) < (4-\epsilon)^t$, for fixed $\epsilon$, has been proved; the best is $r(K_t) < C\frac{\log\log t}{\log t}\binom{2t-2}{t-1}$ claimed by Yackel [**71**]. Now the Erdős-Szekeres bound derives from the inequality $r(k,l) \leq r(k-1,l) + r(k,l-1)$. An examination of this proof reveals that any extremal colouring for $r(k,l)$ on anything approaching $r(k-1,l)+r(k,l-1)$ vertices is jumbled. We can then apply Theorem 6.9 to show the existence of a red $K_k$ or a blue $K_l$, so giving a better upper bound for $r(k,l)$. Certainly this approach gives some improvement over the classical result. At the time of writing, the following theorem at least seems quite likely. The details will be given in [**66**].

**Theorem 6.11.** *There is an absolute positive constant $\epsilon$ such that if $(1-\epsilon)k < l \leq k$ then $r(k,l) < (k+l)^{-\epsilon}\binom{k+l-2}{k-1}$.*

## References


[**1**] M. Ajtai, J. Komlós and E. Szemerédi, Sorting in $C\log n$ parallel steps, *Combinatorica* **3**, 1–19.

[**2**] N. Alon, Eigenvalues and expanders, (preprint).

[**3**] N. Alon, Eigenvalues, geometric expanders, sorting in rounds and ramsey theory, (preprint).

[**4**] N. Alon, Z. Galil and V.D. Milman, Better expanders and superconcentrators, (preprint).

[**5**] N. Alon and V.D. Milman, Eigenvalues, expanders and superconcentrators, in 'Proc. 25th Annual Symp. on Foundations of Computer Science', Florida pp. 320–322.

[6] R. Ayoub, 'An introduction to the analytic theory of numbers', American Mathematical Society (1963).

[7] J.-C. Bermond and B. Bollobás, The diameter of graphs – a survey, in 'Proc. Twelfth Southeastern Conf. on Combinatorics, Graph Theory and Computing', *Congressus Numerantium* **32**, pp. 3–27.

[8] J.-C. Bermond, J. Bond, M. Paoli and C. Peyrat, Graphs and intercommunication networks: diameter and vulnerability, in 'Surveys in Combinatorics' (E.K. Lloyd, ed.) London Math. Soc. Lecture Notes **82** pp. 1–30.

[9] A. Blanchard, quoted by G. Giraud, Nouvelles majorations des nombres de Ramsey binaires-bicolores, *C.R. Acad. Sci. Paris Sér. A* **268** (1969), 5–7.

[10] M. Blum, R.M. Karp, O. Vornberger, C.H. Papadimitriou and M. Yannakakis, The complexity of testing whether a graph is a superconcentrator, *Infor. Process. Letters* **13** (1981), 164–167.

[11] B. Bollobás, 'Extremal Graph Theory', Academic Press, London (1978).

[12] B. Bollobás, 'Random Graphs', Academic Press, London (1985).

[13] B. Bollobás, Geodesics in oriented graphs, *Annals Discrete Math.* **20** (1984), 67–73.

[14] B. Bollobás and P. Catlin, Topological cliques of random graphs, *J. Combinatorial Theory Ser. B.* **30** (1981), 224–227.

[15] B. Bollobás, P. Catlin and P. Erdős, Hadwiger's conjecture is true for almost every graph, *European J. Combinatorics* **1** (1980), 195–199.

[16] B. Bollobás and P. Erdős, An extremal problem of graphs with diameter 2, *Math. Mag.* **48** (1975), 281–283.

[17] B. Bollobás and P. Erdős, Cliques in random graphs, *Math. Proc. Camb. Phil. Soc.* **80** (1976), 419–427.

[18] B. Bollobás and A. Thomason, Graphs which contain all small graphs, *European J. Combinatorics* **2** (1981), 13–15.

[19] B. Bollobás and A. Thomason, Parallel Sorting, *Discrete Appl. Math.* **6** (1983), 1–11.

[20] B. Bollobás and A. Thomason, Random graphs of small order, in 'Random Graphs', *Annals Discrete Math.* (1985), pp. 47–97.

[21] A.E. Brouwer, Some new two-weight codes and strongly regular graphs, *Discrete Applied Math.* **10** (1985), 111-114.

[22] W.G. Brown, On graphs that do not contain a Thomsen graph, *Canad. Math. Bull.* **9** (1966), 281–285.

[23] D.A. Burgess, On character sums and primitive roots, *Proc. London Math. Soc.* **12** (1962), 179–192.

[**24**] N. Calkin, personal communication.

[**25**] F.R.K. Chung, On concentrators, superconcentrators, generalizers and non-blocking networks, *Bell Syst. Tech. J.* **58** (1978), 1765-1777.

[**26**] V. Chvátal and P. Erdős, A note on hamiltonian circuits, *Discrete Math.* **2** (1972), 111-113.

[**27**] M.E. Dyer and A.M. Frieze, Fast solution of some random NP-hard problems (preprint).

[**28**] P. Erdős, Some remarks on the theory of graphs, *Bull. Amer. Math. Soc.* **53** (1947), 292–294.

[**29**] P. Erdős, On the number of complete subgraphs contained in certain graphs, *Publ. Math. Inst. Hung. Acad. Sci., VII, Ser. A* **3** (1962), 459–464.

[**30**] P. Erdős and J.W. Moon, On subgraphs of the complete bipartite graph, *Canad. Math. Bull.* **7** (1964), 35–39.

[**31**] P. Erdős and A. Rényi, On a problem in graph theory, *Publ. Math. Inst. Hungar. Acad. Sci.* **7** (1962), 215–227 (in Hungarian).

[**32**] P. Erdős and J. Spencer, Imbalances in *k*-colorations, *Networks* **1** (1972), 379–385.

[**33**] P. Erdős and G. Szekeres, A combinatorial problem in geometry, *Compositio Math.* **2** (1935), 463–470.

[**34**] R.J. Evans, J.R. Pulham and J. Sheehan, On the number of complete subgraphs contained in certain graphs, *J. Combin. Theory Ser. B* **30** (1981), 364–371.

[**35**] P. Frankl and R.M. Wilson, Intersection theorems with geometric consequences, *Combinatorica* **1** (1981), 357–368.

[**36**] O. Gabber and Z. Galil, Explicit constructions of linear superconcentrators, *J. Comp. and Sys. Sci.* **22** (1981), 407–420.

[**37**] G. Giraud, Sur le problème de Goodman pour les quadrangles et la majoration des nombres de Ramsey, *J. Combin. Theory Ser. B* **27** (1979), 237–253.

[**38**] A.W. Goodman, On sets of acquaintances and strangers at any party, *Amer. Math. Monthly* **66** (1959), 778–783.

[**39**] R.L. Graham and J.H. Spencer, A constructive solution to a tournament problem, *Canad. Math. Bull.* **14** (1971), 45–48.

[**40**] G.R. Grimmett and C.J.H. McDiarmid, On colouring random graphs, *Math. Proc. Cambridge Phil. Soc.* **77** (1975), 313–324.

[**41**] A. Gyárfás, J. Lehel, R.H. Schelp and Zs. Tuza, Ramsey numbers for local colorings, (preprint).

[**42**] A. Gyárfás, J. Lehel, J. Nešetřil, V. Rödl, R.H. Schelp and Zs. Tuza, Local *k*-colorings of graphs and hypergraphs, (preprint).

[43] R. Häggkvist and P. Hell, Parallel sorting with constant time for comparisons, *SIAM J. Comput.* **10** (1981), 465–472.

[44] J. Haviland and A. Thomason, Pseudo-random hypergraphs (to appear).

[45] X.L. Hubaut, Strongly regular graphs, *Discrete Math.* **13** (1975), 357–381.

[46] Sh. Jimbo and A. Maruoka, Expanders obtained from affine transformations, in 'Proc. 17th Annual ACM Symp. on Theory of Computing' (1985), pp. 88–97.

[47] N.M. Katz, 'Sommes exponentielles', *astérisque* **79**, Société Mathématique de France (1980).

[48] A. Kostochka, A lower bound for the Hadwiger number of graphs by their average degree, *Combinatorica* **4** (1984), 307–316.

[49] P. Kövári, V.T. Sós and P. Turán, On a problem of K. Zarankiewicz, *Colloq. Mat.* **3** (1954), 50–57.

[50] G. Lorden, Blue-empty chromatic graphs, *Amer. Math. Monthly* **69** (1962), 114–120.

[51] W. Mader, Homomorphiesätze für Graphen, *Math. Ann.* **178** (1968), 154–168.

[52] G.A. Margulis, Explicit constructions of concentrators, *Problemy Peredachi Informatsii* **9**(4) (1973), 71–80 (in Russian). English translation in *Problems Info. Transmission*, Plenum Press (1975), 325–332.

[53] R. Mathon, Lower bounds for ramsey numbers and assosciation schemes, *J. Combinatorial Theory, Ser. B* **42** (1987), 122–127.

[54] D.W. Matula, On the complete subgraph of a random graph, in 'Combinatory Mathematics and its Applications', Chapel Hill, N.C., pp. 356–369.

[55] H.L. Montgomery, Topics in multiplicative number theory. *Lecture Notes in Mathematics* **227**, Springer-Verlag (1971).

[56] R.G.E. Pinch, A sequence well distributed in the square, *Math. Proc. Cambridge Phil. Soc.* **99** (1986), 19–22.

[57] J.J. Seidel, A survey of two-graphs, in 'Colloquio Internazionale sulle Teorie Combinatorie', Atti dei Convegni Lincei **17**, Accad. Naz. Lincei, Roma (1976), pp. 481–511.

[58] L.A. Székely, On the number of homogeneous subgraphs of a graph, *Combinatorica* **4** (1984), 363–372.

[59] R.M. Tanner, Explicit construction of concentrators from generalized $N$-gons, *SIAM J. Alg. Discr. Meth.* **5** (1985), 287–293.

[60] A. Thomason, On finite ramsey numbers, *European J. Combinatorics* **3** (1982), 263–273.

[61] A. Thomason, An extremal function for contractions of graphs, *Math. Proc. Cambridge Phil. Soc.* **95** (1984), 261–265.

[62] A. Thomason, Pseudo-random graphs, in 'Proceedings of Random Graphs, Poznań 1985', (M. Karonski, ed.) *Annals of Discrete Math.* (1987).

[63] A. Thomason, A disproof of a conjecture of Erdős in ramsey theory, (submitted)

[64] A. Thomason, A linear expected time hamilton cycle algorithm, (submitted).

[65] A. Thomason, Dense expanders, (to appear).

[66] A. Thomason, Upper bounds for ramsey numbers, (to appear).

[67] L.G. Valiant, Graph theoretic properties in computational complexity, *J. Comp. and Sys. Sci.* **13** (1976), 278–285.

[68] W.F. de la Vega, On the maximum density of graphs which have no subcontraction to $K^s$, *Discrete Math.* **46** (1983), 109–110.

[69] K. Walker, Dichromatic graphs and ramsey numbers, *J. Combinatorial Theory* **5** (1968), 238–243.

[70] A. Weil, Sur les courbes algébrique et les variétés qui s'en déduisent, *Actualités Sci. Ind. No. 1041* (1948).

[71] J. Yackel, Inequalities and asymptotic bounds for ramsey numbers, *J. Combinatorial Theory Ser. B* **13** (1972), 56–68.

[72] Š. Znám, On a combinatorial problem of K. Zarankiewicz, *Colloq. Math.* **11** (1963), 81–84.

# THE METRIC STRUCTURE OF GRAPHS: Theory and Applications

Peter Winkler
Emory University, Atlanta, Georgia 30322 U.S.A.

*Abstract.* The "path metric" on the vertices of a connected graph $G$ is given by defining the distance between two vertices $u$ and $v$ to be the minimum over all paths $P$ from $u$ to $v$ of the number of edges in $P$. Over the past 20 years a structure theory has emerged whose aim is to simplify the path metric by means of isometric embedding in cartesian products of graphs. Although the theory is peculiar to graphs, it looks like a typical algebraic structure theory and the analog of representation by subdirect products works out nicely.

Development of the theory was motivated by a problem from computer science and has contributed to the solution of several others. These problems, which arise in complexity theory, network design and data structures, will be described briefly and their connections with the structure theory explained.

## Introduction

The general aim of a structure theory is to break down a complex object into simpler components, the object in this case being the path metric on a graph. The approach taken is a familiar one to algebraists, namely representation by embedding in a product; it turns out that this approach works quite well here, despite the apparent absence of category-theoretic underpinnings. There are a variety of applications of the structure theory to computer science, some of which will be briefly presented below.

By a *graph* $G = \langle V(G), E(G) \rangle$ we shall mean a finite set $V(G)$ of *vertices* together with a collection $E(G)$ of unordered pairs of vertices called *edges*, that is, a finite, simple graph. For definition of basic terms such as "path" or "connected" the reader is referred to any elementary text on graph theory or combinatorics. The *distance* $d_G(u,v)$ between two vertices of a connected graph $G$ is defined to be the number of edges in a shortest path from $u$ to $v$; the resulting "path metric" turns $V(G)$ into a metric space. An *isometric embedding* of $G$ in another graph $H$ is a map $\alpha: V(G) \to V(H)$ such that $d_H(\alpha(u),\alpha(v)) = d_G(u,v)$ for all vertices $u$ and $v$ of $G$; the image of $G$ is then a subgraph of $H$ isomorphic to $G$.

**The Cartesian Product**

There are at least three natural graph products definable on the cross product of the vertex sets of two graphs; a clever notation scheme due to J. Nešetřil assigns a symbol to each which looks like the corresponding product of two edges. Although the "weak product" (denoted by $\times$) and the "strong product" ($\boxtimes$) are universal products in the category of graphs (as non-reflexive and reflexive relational structures, respectively), the "cartesian" product ($\square$) has turned out to be extremely versatile and is the product of choice in metric theory, for reasons which will become plain.

The cartesian product $G\square H$ of graphs $G$ and $H$ is defined as follows. The vertices of $G\square H$ are ordered pairs $(g,h)$ with $g\in V(G)$ and $h\in V(H)$, and $\{(g,h),(g',h')\}$ is an edge of $G\square H$ iff *either* $\{g,g'\}$ is an edge of $G$ and $h=h'$, *or* $\{h,h'\}$ is an edge of $H$ and $g=g'$.

It is easily seen that the cartesian product is associative and that $\{(g_1,g_2,\ldots,g_k),(g_1',g_2',\ldots,g_k')\}$ is an edge of $G_1\square G_2\square\cdots\square G_k$ just when $\{g_i,g_i'\}$ is an edge of $G_i$ for some $i$ and $g_j=g_j'$ for all $j\neq i$, in which case we say that the edge "belongs to the $i$th factor." Note that the product of $k$ copies of $K_2$, which we denote by $K_2^k$, is just the "Boolean $k$-cube," i.e. the graph given by the vertices and edges of a (unit) cube in $k$-space; its metric is of course the $k$-dimensional Hamming metric.

The cartesian product of $K_3$ and the three-point path $P_3$ is illustrated in Fig. 1 below.

Sabidussi (1960), and later Vizing (1963), showed that the cartesian product enjoys unique prime factorization for connected graphs. The proofs were difficult and did not provide an efficient method for *finding* the prime factorization---a point which we will address again later. Unique prime factorization fails for disconnected graphs, for which Zaretskii (1965) gave the following example: $K_2^5+K_2^4+K_2^3+K_2^2+K_2+K_1 = (K_2^3+K_1)\square(K_2^2+K_2+K_1) = (K_2+K_1)\square(K_2^4+K_2^2+K_1)$, where "+" denotes disjoint union and all the graphs in parentheses can be shown to be prime.

Since each step of a path in a product $\prod G_i$ moves in only one factor, any path in a product breaks down into paths in the factors. Thus we have that

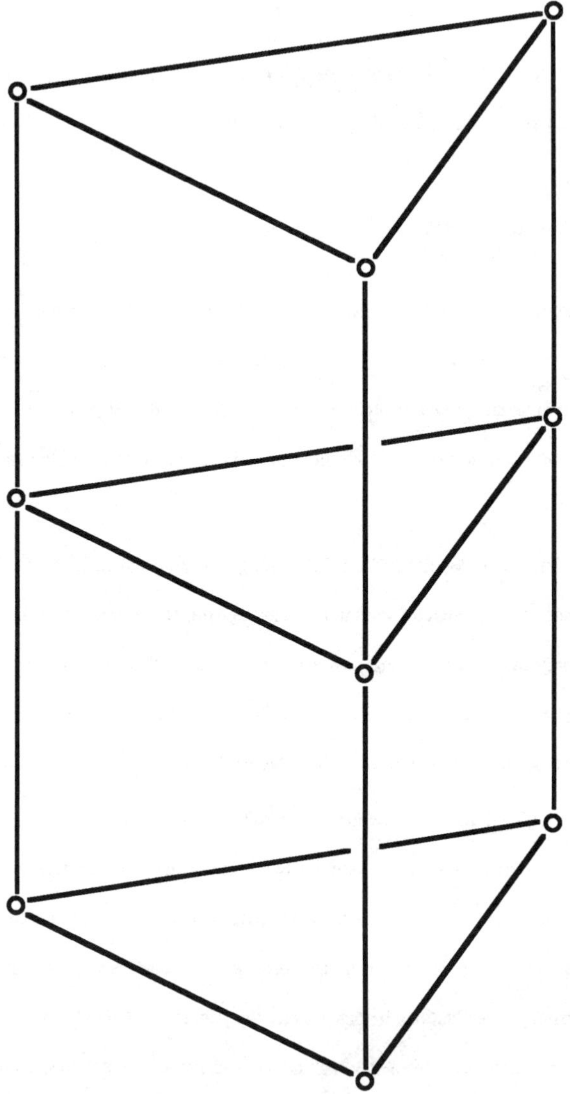

Fig. 1

$$d_{\prod G_i}((g_1,\ldots,g_k),(g_1',\ldots,g_k')) = \sum d_{G_i}(g_i,g_i')\,.$$

It follows that if $G$ is a product of simpler graphs, *or if it is an isometric subraph of such a product,* then its metric can be broken down into simpler components.

**Isometric Embedding in Products**

In representing the metric on a graph $G$, the best result one can reasonably hope to encounter is an isometric embedding of $G$ in $K_2^k$; equivalently, a description of $G$ as a subspace of Hamming space. Such embeddings were sought by Firsov (1965) for the purpose of investigating "the closeness of various linguistic objects," but only preliminary results were obtained at that time.

In the early '70's Graham & Pollak (1971, 1972) highlighted the problem in the process of designing addressing schemes for computer communications networks. Information is sent from computers in "packets" only microseconds long, and efficiency requires that many of these be on a network simultaneously. Each packet is augmented with the address of its destination, and must somehow find a short route to that destination through the network.

Graham and Pollak's idea was to represent the network by a graph and provide addresses for the vertices in such a way that the distance between two vertices (number of edges in a shortest path) can be easily recovered by comparing their addresses. Then when a packet reaches a vertex it can compare its destination address with the local address, discovering, say, that it is presently at distance $k$ from its destination. It then checks the addresses of neighboring vertices until it finds one at distance only $k-1$ from the destination, and moves there. Thus using local information only, it finds a shortest path to its destination.

The most obvious way to realize Graham and Pollak's objective is to address the vertices by binary strings whose Hamming distances match the graph distances; that is, to embed the graph isometrically in a Boolean cube. Unfortunately, as noted already by Firsov, there are graphs which are resistant to such embeddings even when made bipartite by edge-subdivision. Graham and Pollak then turned to the addition of a "don't care" symbol ("*") to the binary strings, whose distances to the other symbols 1 and 0 are both defined to be zero. Even though the space of all such strings (of a given length) is no longer a metric space, the new scheme worked beautifully; later Winkler (1983) was able to show that any graph on $n$ vertices could be addressed by strings of this sort having length at most $n-1$.

The question of which graphs could be addressed by the $\{0,1\}$ scheme, however, remained open until Djoković (1973) gave a complete characterization. If $\{x,y\}$ is an edge of a connected graph $G$, let $Nxy = \{v \in V(G): d_G(v,x) = d_G(v,y)-1\}$, i.e. the set of vertices nearer to $x$ than to $y$, and $xNy = \{v \in V(G): d_G(v,x) = d_G(v,y)\}$. Further, any set $S$ of vertices of $G$ is deemed *convex* if every shortest path in $G$ between vertices of $S$ is contained completely in the subgraph induced by $S$.

**Theorem** (Djoković 1973). A connected graph $G$ can be isometrically embedded in a Boolean cube if and only if $G$ is bipartite and for every edge $\{x,y\}$ of $G$, both $Nxy$ and $Nyx$ are convex.

We omit the proof of Djoković's theorem since it will follow from more general results below.

If more symbols are allowed in the addressing scheme, with the distance between distinct symbols still always 1, then the problem becomes one of embedding $G$ isometrically in some power $K_m^k$ of a complete graph. This also cannot generally be done, as can be seen for example by examining the complete bipartite graph $K_{2,3}$ or the cycle $C_5$. A polynomial-time characterization of the graphs which can be embedded in a power of a complete graph (equivalently, in a product of complete graphs) was obtained by Winkler (1984), who also made the (then) surprising observation that *all* irredundant isometric embeddings of this sort were unique up to symmetries of the product graph.

By 1984 it was beginning to look like some much more general pleasantnesses concerning isometric embedding in products lay behind the previous results. Let us define a *metric representation* of a connected graph $G$ to be an isometric embedding of $G$ in a product $H_1 \square \cdots \square H_k$

which is "irredundant" in the following sense: (1) each $H_i$ is a connected graph with at least two vertices, and (2) every vertex of each factor appears as a coordinate in the image of at least one vertex of $G$. (It is not hard to show that any isometric embedding in a product can be made irredundant by discarding unused vertices and trivial factors.)

Two (metric) representations are *equivalent* if there is a bijection between the factors of one and the factors of the other, together with isomorphisms between corresponding factors for which the obvious diagram commutes. A graph $G$ is said to be *irreducible* if all of its representations are trivial, that is, equivalent to the identity map from $G$ to itself. All of this is of course a parody, in a sense, of subdirect products in algebra, but as far as we know no algebraic interpretation is available. Nonetheless everything one could hope to be true actually is.

**Theorem** (Graham & Winkler (1984,1985)). Every connected graph $G$ has a unique *canonical* representation $G \to G_1 \square \cdots \square G_k$ in which every factor is irreducible. For any other representation $G \to H_1 \square \cdots \square H_m$, there is a surjection $\phi$: $\{1,...,k\} \to \{1,...,m\}$ between the index sets and representations $H_i \to \prod\{G_j: \phi(j)=i\}$ for which everything commutes; that is, the canonical representation can be factored through any other. Furthermore, the canonical representation of $G$ can be obtained by a polynomial-time algorithm.

**Proof:** Fix a representation $G \to H_1 \square \cdots \square H_m$ of $G$ and denote by $(v_1, \ldots, v_m)$ the image of a vertex $v$ of $G$. Let $e=\{x,y\}$ be an edge of $G$ with $Nxy$, $xNy$ and $Nyx$ (defined above) partitioning $V(G)$. (See Fig. 2.) If $e$ belongs to the $i$th factor, that is, if $x_i \neq y_i$, then of course $x_j=y_j$ for all $j \neq i$ so for any vertex $v$,

$$d_G(v,x)-d_G(v,y) = d_{H_i}(v_i,x_i)-d_{H_i}(v_i,y_i) .$$

It follows that the location of $v$ in the above partition depends only on $v_i$. If $f=\{u,v\}$ is an edge which crosses the partition, e.g. $u \in Nxy$ and $v \in xNy$, then $u_i \neq v_i$ and therefore $f$ and $e$ belong to the same factor.

Define the relation $\theta$ on $E(G)$ as follows: if $e=\{x,y\}$ and $f=\{u,v\}$ are edges then

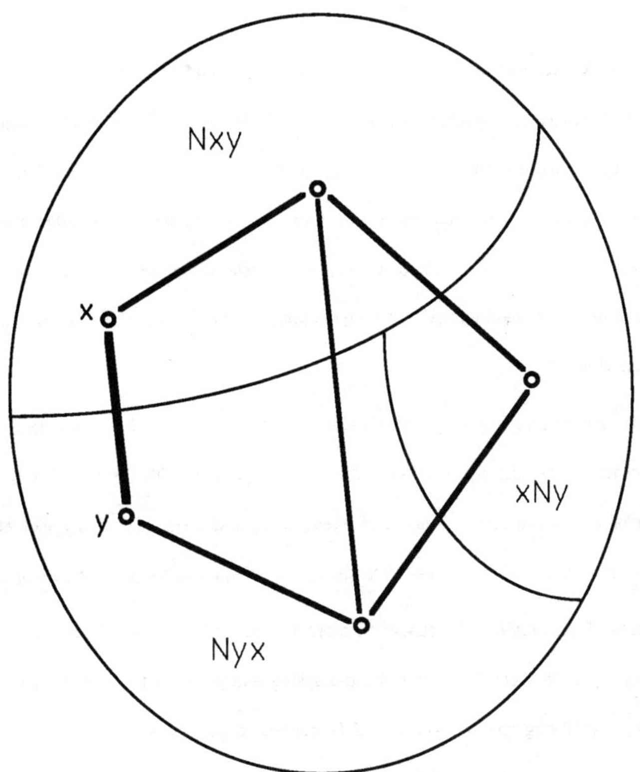


Fig. 2

$$e\theta f \text{ iff } d(x,u)+d(y,v) \neq d(x,v)+d(y,u) \,.$$

In that case $f$ indeed crosses the partition induced by $e$, so $e$ and $f$ belong to the same factor. The relation $\theta$ is reflexive and symmetric but not generally transitive; let $\hat{\theta}$ be its transitive closure. Then $e\,\hat{\theta} f$ again implies that $e$ and $f$ belong to the same factor. If the equivalence classes of $\hat{\theta}$ in $E(G)$ are $E_1, E_2, \ldots, E_k$ then we already know that $m \leq k$, and in particular if $k=1$ then $G$ is irreducible. The surprise is that there actually is a representation with $k$ factors, corresponding to the rather *ad hoc*-looking relation $\hat{\theta}$. The number $k$ of classes of $\hat{\theta}$ will be called the metric *dimension* $\dim(G)$.

Computing the relation $\hat{\theta}$ is easy enough since the distance matrix is obtainable in polynomial time. In practice one can often obtain $\hat{\theta}$ by hand, beginning with the fact that $\hat{\theta}$ matches opposite edges of any isometrically embedded even cycle, and all edges of any isometric odd cycle. In Fig. 3 below the three equivalence classes of $\hat{\theta}$ are indicated by lines of various textures.

The process of determining the canonical representation of $G$ from $\hat{\theta}$ is easy and, at each step, forced. Let $G_j$ be the graph obtained by collapsing every edge of $G$ which is *not* in the class $E_j$, thereby identifying any two vertices of $G$ connected by a path none of whose edges lie in $E_j$. This defines also a natural map $\sigma_j$ from $V(G)$ to $V(G_j)$, and the map $\sigma \colon G \to G_1 \Box \cdots \Box G_k$ given by

$$\sigma(v) = (\sigma_1(v), \ldots, \sigma_k(v))$$

is asserted to be our canonical representation. The canonical representation of the graph of Fig. 3 is shown by means of vertex labelling in Fig. 4, and pictorially in Fig. 5 as an isometric subgraph of the product.

To show that $\sigma$ is, in fact, an isometric embedding, the following critical lemma is required. Its proof actually contains the crux of the proof of the theorem.

**Lemma.** Let $P$ be a shortest path between vertices $x$ and $y$ of $G$, and let $N_j$ be the number of edges of $P$ which belong to the class $E_j$. Then every other path from $x$ to $y$ contains at least $N_j$ edges of the class $E_j$.

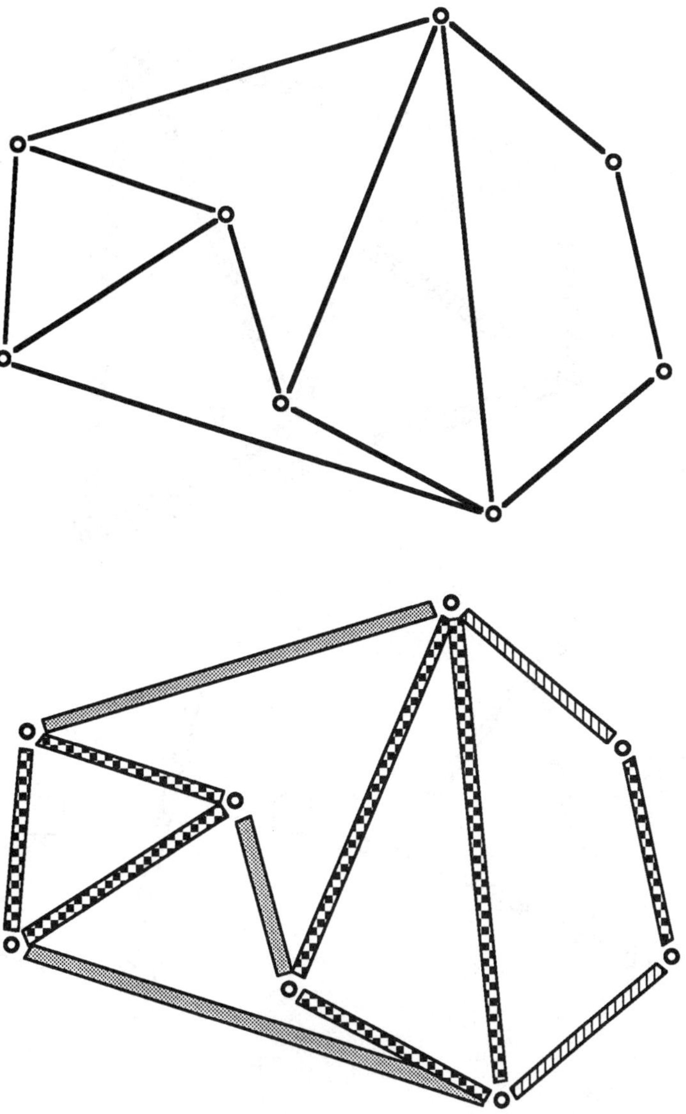

Fig. 3

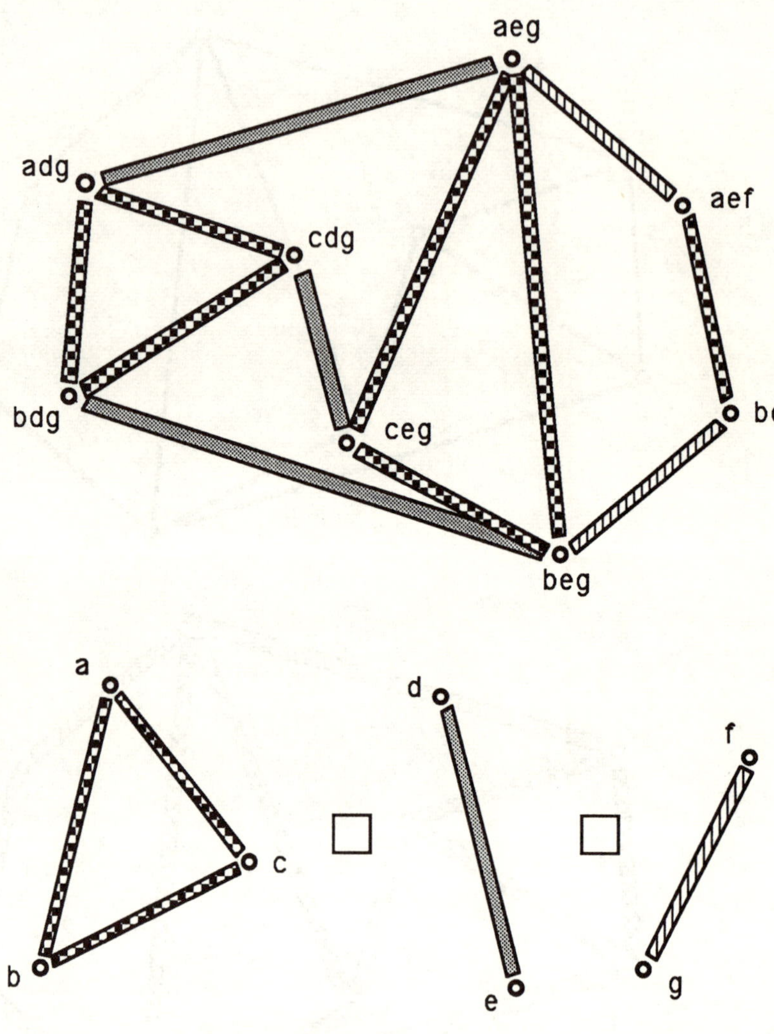


Fig. 4

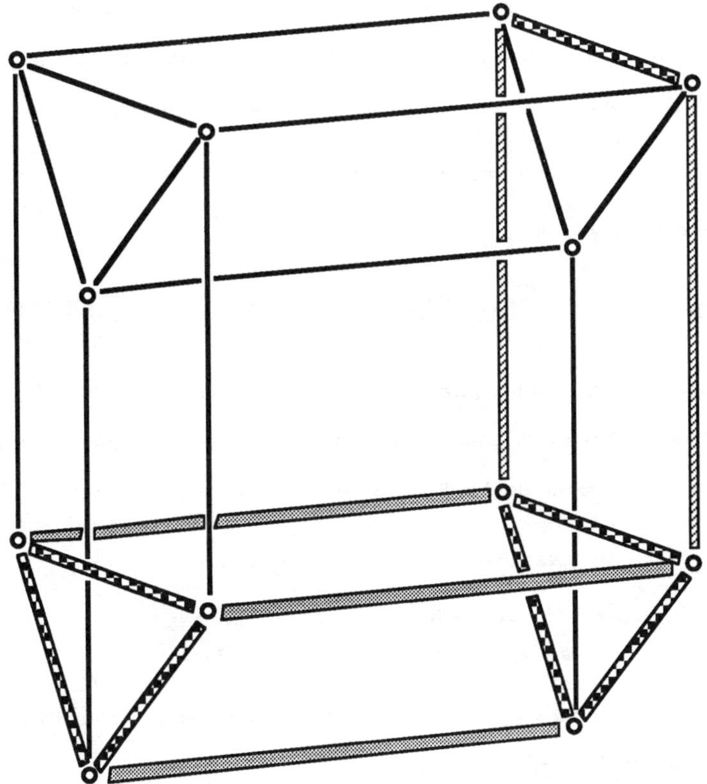

Fig. 5

*Proof of Lemma:* Let $P$ be given by vertices $z(0),z(1),\ldots,z(d)$ where $x=z(0)$, $y=z(d)$ and $d=d_G(x,y)$. It will be useful to define some auxilliary functions from $V(G)$ to the integers. For each $i$, $1\le i\le d$, let $a_i(u)=d(u,z(i))-d(u,z(i-1))$; and for each $j$, $1\le j\le k$, set

$$b_j(u)=\sum\{a_i(u):\{z(i-1),z(i)\}\in E_j\}\,.$$

It follows that for any vertex $v$,

$$\sum_{i=1}^{d}a_i(v)=\sum_{j=1}^{k}b_j(v)=d(v,y)-d(v,x)\,.$$

Now suppose that $\{u,v\}$ is an edge belonging to $E_{j'}$; then $d(z(i-1),u)+d(z(i),v)=d(z(i),u)+d(z(i-1),v)$ whenever the edge $\{z(i-1),z(i)\}$ is *not* in $E_{j'}$, since the two edges are not related by θ. Hence $b_j(u)=b_j(v)$ for all $j\ne j'$. Furthermore,

$$\begin{aligned}|b_{j'}(u)-b_{j'}(v)| &= |\sum_{j=1}^{k}b_j(u)-\sum_{j=1}^{k}b_j(v)|\\ &= |(d(u,y)-d(u,x))-(d(v,y)-d(v,x))|\\ &\le |d(u,y)-d(v,y)|+|d(u,x)-d(v,x)|\le 1+1=2\,.\end{aligned}$$

We have thus shown that along an edge of $E_{j'}$ the function $b_j$ does not change for $j\ne j'$, and changes by at most 2 for $j=j'$.

Now let $Q$ be any path from $x$ to $y$ and fix $j'$ between 1 and $k$. Since $P$ is a shortest path, we have of course that $a_i(x)=1$ and $a_i(y)=-1$ for each $i$, $1\le i\le d$; it follows that $b_{j'}(x)=N_{j'}$ and $b_{j'}(y)=-N_{j'}$. But then, moving along the path $Q$ from $x$ to $y$, the function $b_{j'}$ must change in value by $2N_{j'}$. Since all changes take place on edges of $Q\cap E_{j'}$, and none of these changes are by more than 2, the number of edges of $Q$ which belong to $E_{j'}$ must be at least $N_{j'}$, proving the Lemma.

To show that σ is isometric, let $x$ and $y$ be arbitrary vertices of $G$ and let $H$ stand for the target product $\prod G_j$. Then

$$d_H(\sigma(x),\sigma(y))=\sum_{j=1}^{k}d_{G_j}(\sigma(x),\sigma(y)).$$

By definition of the factor graphs $G_j$, however, $d_{G_j}(\sigma(x),\sigma(y)$ is merely the minimum over all paths $Q$ from $x$ to $y$ of the number of edges of $Q$ which lie in $E_j$. Since by the Lemma that minimum is attained in any *shortest* path $P$ from $x$ to $y$ by the number $N_j$, we have that

$$d_H(\sigma(x),\sigma(y)) = \sum_{j=1}^{k} N_j = d_G(x,y)$$

as required.

The Lemma shows also that the endpoints of an edge in $E_j$ are not identified in the formation of $G_j$, and therefore the canonical factors all have at least two vertices; thus, since the collapsing maps $\sigma_j$ are surjective, the map $\sigma$ is irredundant. That the factors are irreducible is clear, since if $G_j$ had a non-trivial representation, its target product could be substituted for $G_j$ to give a representation of $G$ with more than $k$ factors.

Any factor $H_i$ of another representation of $G$ must "contain" exactly the edges in $\bigcup\{E_j\colon j \in J\}$ for some non-empty set $J$ of indices, and will be isomorphic to the graph obtained by collapsing all edges not in that union; it is not hard to see that the factors of the canonical representation of $H_i$ itself are isomorphic to the graphs $G_j$ for $j \in J$, and that everything commutes as claimed. In fact, there is exactly one representation of $G$ (up to equivalence) for every partitioning of the index set $\{1,2, \ldots ,k\}$. Thus the number of equivalence classes of metric representations of a graph $G$ is the Bell number $\mathbf{B}(\dim(G))$.

It remains only to note that, since the $\dim(G)$ is at most the number of edges of $G$, the process described above constitutes a polynomial-time algorithm for computing the canonical representation of $G$. Caution: the size of the target graph $\prod G_j$ may be exponential in $|G|$. Fortunately, description of the representation does not require listing the vertices of $\prod G_j$ but only specifying the factor graphs and associating $k$-tuples of their vertices to vertices of $G$ as in Fig. 4. □

At this point it seems worthwhile to step back and look at some special cases. When $G$ is a tree on $n$ vertices, the $\hat{\theta}$ relation is just the diagonal; hence each $E_j$ is a single edge and $G$

embeds into $K_2^{n-1}$. In fact, among graphs on $n$ vertices only a tree can have dimension as high as $n-1$. To see this note first that by the Lemma, any spanning tree of a graph $G$ contains at least one edge of every $\hat{\theta}$-class (since it contains a path between any two vertices). Thus dim($G$) is at most $k-1$, so that if $G$ is not a tree, there are two edges in the same class; a spanning tree which contains both of those edges can cover only $k-2$ classes, so dim($G$) is strictly less than $k-1$.

In an even cycle $C_{2n}$, $\hat{\theta}$ associates antipodal edges, resulting in representation in $K_2^n$. Complete graphs are irreducible, justifying the earlier statement that isometric embeddings in products of complete graphs are unique up to symmetries of the product. Odd cycles are also irreducible (closure of $\theta$ under transitivity pulls in all the edges) as are "most" graphs in all models of random graphs that we know of. Graphs which are themselves cartesian products will have at least as many factors in the canonical representation as in the product, with equality when the product factors are irreducible.

Let $G$ be arbitrary and fix an edge $\{x,y\}$ of the class $E_j$. If $E_j$ consists *only* of edges which travel between parts of the partition $V(G) = Nxy \cup xNy \cup Nyx$ then each non-empty part will collapse into a single point in the formation of the canonical factor $G_j$, resulting in $G_j \cong K_2$ when $xNy$ is empty and $G_j \cong K_3$ otherwise. Conversely if $G_j$ is isomorphic to $K_2$ or $K_3$ then no edges can have been added to $E_j$ in passing from $\theta$ to $\hat{\theta}$. Since $K_2$ and $K_3$ are the only nontrivial graphs isometrically embeddable in $K_3$, we have:

**Corollary 1.** $G$ is isometrically embeddable in a power of $K_3$ if and only if the relation $\theta$ on $E(G)$ is transitive.

If $G$ is bipartite then the class $xNy$ is always empty for any edge $\{x,y\}$, since otherwise an odd cycle would result; conversely any minimum-sized odd cycle is isometric and produces a non-empty $xNy$ for any of its edges. Thus we also have:

**Corollary 2.** $G$ is isometrically embeddable in a hypercube if and only if $G$ is bipartite and the relation $\theta$ on $E(G)$ is transitive.

If in a bipartite graph $G$ one of the $Nxy$'s fails to be convex, then the edge $\{x,y\}$ together with the two crossing edges of the offending path witness a failure of transitivity for $\theta$; conversely if $e\,\theta\{x,y\}\theta f\,\theta e$ then $e$ and $f$ lie on a geodesic testifying to the failure of convexity of either $Nxy$ or $Nyx$. Djoković's Theorem (above) follows. The argument does not generalize to non-bipartite graphs, where two edges not in the $\theta$ relation need not both lie on some geodesic.

## Factoring

In the 'sixties (after Sabidussi's (1960) proof of unique factorization for the cartesian product of graphs) Karp and Cook developed the theory of polynomial-time algorithms and NP-completeness, for which we refer the reader to Garey & Johnson (1979). The problem of factoring a graph was a natural candidate for complexity analysis, partly because of interest in the complexity of factoring *numbers* and testing primality, but later mostly because while many graph problems were falling into the class "Polynomial" or the class "NP-complete," the factoring problem resisted classification. Welsh (1982) formally posed the problem of determining the complexity of graph factorization, suggesting that by analogy with the number case, perhaps factoring would be the first example in graph theory of a problem which was in NP and in co-NP, but not in P.

The most natural statement of the factoring problem, for the purposes of complexity analysis, is the following:

FACTORING

*Instance.* (The adjacency matrix of) a connected graph $G$.

*Question.* Is $G$ composite, that is, can $G$ be expressed as a non-trivial cartesian product?

Clearly FACTORING is in NP, since the two (or more) factor graphs would themselves be a polynomial certificate for a "yes" answer. For any fixed connected graph $H$, the following problem is also in NP:

FACTORING BY $H$

*Instance.* A connected graph $G$.

*Question.* Is there a graph $J$ such that $G \cong H \square J$ ?

One might also ask whether given $G$ and $H$, $H$ divides $G$; however, this is not necessarily in NP because even if the prime factorization of $G$ were known, the determination of whether one of the factors is isomorphic to $H$ (the "GRAPH ISOMORPHISM" problem---see Garey & Johnson (1979)) is not (yet) known to be possible in polynomial time. GRAPH ISOMORPHISM interferes also when trying to factor a *dis*connected graph.

It is perhaps worth remarking that even though factoring a graph on $n$ vertices entails factoring the number $n$, factoring a graph is not necessarily as hard as factoring a number (from the point of view of complexity theory). The reason is that when the instance is an $n$-vertex graph one is "credited" with input length $n^2$, whereas the input length of a number is only its logarithm base 2.

The classification of FACTORING and FACTORING BY $H$ was finally done in 1984, independently by Feigenbaum et. al. (1985) and Winkler (1986). The result, happily, is that both are polynomial. Feigenbaum, Hershberger and Schäffer succeeded in showing that Sabidussi's original approach could be converted to a polynomial algorithm (time $O(n^{4.5})$); the proof is quite difficult (not surprisingly, since otherwise it might have been done long before). The present author used metric structure theory, and the proof is given below; it requires little more than the machinery developed above.

We begin, of course, by obtaining the canonical metric representation $\sigma: G \to \prod G_j$ of the input graph $G$, where the number of canonical factors is $k = \dim(G)$. Since the isomorphism $\alpha: G \to H_1 \square \cdots \square H_m$, where $\prod H_i$ is the unique prime factorization of $G$, is also a representation, it will induce a partition $(S_1, \ldots, S_m)$ of the canonical index set $S = \{1, \ldots, k\}$. Each cartesian factor $H_i$ will have $\{G_j : j \in S_i\}$ as its own set of canonical factors.

If a representation is given, it is easy to check if it is a factorization (for example, by comparing the product of the orders of the factors to the order of $G$). The difficulty is that the number of representations of $G$ (i.e., number of partitions of $S$) is exponential in $k$, which could be of the same order of magnitude as the number $n$ of vertices of $G$.

Fix a subset $T$ of $S$, let $H_T$ be the product of the canonical factors $G_j$ for $j \in T$, and let $\sigma_T: G \to H_T \,\square\, \prod_{j \in T} G_j$ be the map $\sigma$ with codomain regarded as an $(n-|T|+1)$-fold product. If $\sigma_T$ is irredundant (equivalently, is a representation) then the set $T$ will be said to be "complete." To check whether $T$ is complete is a polynomial operation: if $H_T$ has more than $n$ vertices than $T$ cannot be complete; otherwise we may simply check to see if every vertex of $H_T$ occurs in the range of $\sigma_T$.

Every singleton $\{j\}$ is complete since $\sigma$ itself is irredundant, and every subset of a complete set is complete. The full index set $S$ will be complete just when $\sigma$ is already surjective, i.e. when the canonical representation is the prime factorization. If $S$ is incomplete then we can find in polynomial time a subset $T$ of $S$ which is *minimally* incomplete, i.e. $T$ is incomplete but every proper subset of $T$ is complete. To do this we look for a complete set $S'$ of size $k-1$; if none exists, $S$ is itself minimally incomplete. Otherwise we check the subsets of size $k-2$ of $S'$, etc.

We claim that $T$ is contained in one of the sets $S_j$ of the partition defined by the factoring of $G$. If not, let $T_1, \ldots, T_q$ be the non-empty intersections of $T$ with various $S_i$'s; since each $T_i$ is complete and $\alpha$ is surjective, $T = \bigcup T_i$ must also be complete, a contradiction.

It follows that if $T$ is minimally incomplete, then the representation given by making $\sigma_T$ irredundant (by casting out unused vertices of $H_T$) still factors through $\alpha$. We now repeat the whole procedure with the new representation, which has fewer factors. Eventually we either reduce to a single factor (in which case $G$ is prime) or to a representation with complete index set, in which case that representation is the unique prime factorization of $G$.

We are indebted to J. Feigenbaum and A.A. Schäffer for pointing out that the above algorithm can, with care, be done in time $O(n^4)$ in the worst case.

Fig. 6 below illustrates the simplest non-trivial example of the above factoring procedure that we could find. The graph pictured has three copies of $K_2$ as its canonical factors, with vertices labelled $\{a,b\}$, $\{c,d\}$, and $\{e,f\}$. Since no vertex of $G$ is mapped to a tuple beginning with "$(b,d,\ldots,$" the subset {1,2} of the canonical index set {1,2,3} is incomplete (perforce minimally). The first two canonical factors are thus replaced by an isometric subgraph of their product---namely, the path $P_3$---whose vertices we may as well denote by $ad$, $ac$ and $bc$. The new realization is given in the lower part of the figure, and since the full (two-element) index set is now complete, it is a genuine factorization.

Walker (1986) has extended this procedure to polynomial algorithms for factoring directed graphs and even partially ordered sets (using the fact that the ordinary relational product on partially ordered sets corresponds to the cartesian product of their Hasse diagrams). Walker has also used the metric represention theory to show that any two cartesian factorizations of a connected (but not necessarily finite) graph have a strict common refinement, a strengthening of Sabidussi's unique factorization theorem. Imrich (1986) has used the theory both to reprove Sabidussi's theorem and to investigate the behavior of automorphism groups of cartesian products. Since none of these questions seem at first to have anything to do with metrics, we think that the value of the metric approach is particularly well documented here.

## Games and Retracts

Structure theories different from the one described above have been built using the concept of "retract" in place of our "isometric subgraph." (We refer the reader for example to Hell (1976) or Nowakowski & Rival (1982)). When combined with the cartesian product, the retract turns out to be particularly suitable for the study of games on graphs (as we shall see) and is closely tied to metrics.

By a graph *homomorphism* $G \to H$ we shall mean a map $\phi$: $V(G) \to V(H)$ such that for any

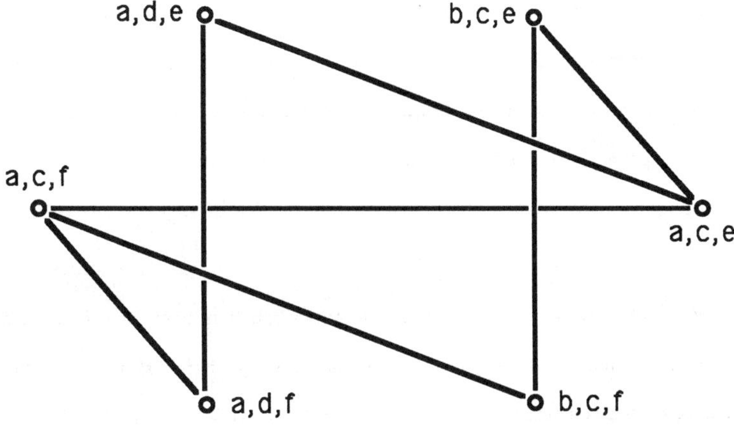


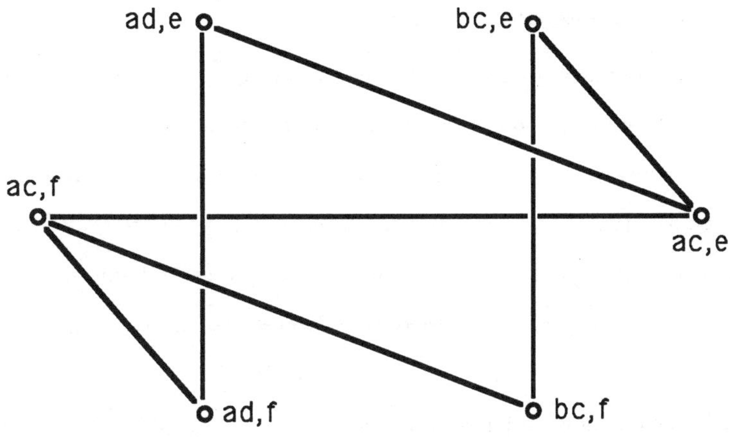


Fig. 6

edge $\{x,y\}$ of $G$, either $\phi(x)=\phi(y)$, or $\{\phi(x),\phi(y))\}$ is an edge of $H$. (This is sometimes called a "weak" homomorphism; in a "strong" homomorphism an edge cannot be contracted to a point.) An equivalent defining condition is that $d_H(\phi(u),\phi(v)) \leq d_G(u,v)$ for any two vertices $u,v$ of $G$; thus a homomorphism constitutes a "contraction" (in the weak sense) from a metric point of view. The maps $\sigma_j: G \rightarrow G_j$ defining the canonical representation of $G$ are examples of homomorphisms.

A (weak) *retraction* $\rho$ of a graph $G$ onto a subgraph $H$ is a homomorphism from $G$ to $H$ which is the identity on $H$, i.e. such that $\rho(x)=x$ for all $x \in V(H)$. $H$ is said to be a (weak) *retract* of $G$ if it is a subgraph of $G$ for which such a retraction exists. Note that since $\rho$ contracts distances, and distances cannot be smaller in $H$ than in $G$, every retract is necessarily an isometric subgraph. The converse fails, e.g. the cycle $C_6$ can be isometrically embedded in $K_2^3$ via the canonical representation, but no retraction exists.

In a game played on a graph it is frequently the case that a strategy which works on graphs $G_1, \ldots, G_k$ can be made to work on the cartesian product $G_1\Box \cdots \Box G_k$ by "playing on all factors simultaneously"; similarly, strategies can often be mapped down from a graph to its retract. One example (another will be seen shortly) is the pursuit game of Nowakowski & Winkler (1983), in which Player I (the "cop") chooses a vertex of some fixed graph $G$, Player II (the "robber") chooses another, and the players move alternately thereafter beginning with the cop. A move consists of either staying at one's present vertex or moving to an adjacent one, and the cop wins if he can catch the robber in a finite number of moves. A graph is called "cop-win" if the cop has a winning strategy; it is easily seen that any retract of a product of cop-win graphs is cop-win.

An obvious question to ask is the analogue of Djoković's theorem: which graphs are retracts of hypercubes? In a connected graph $G$, the vertex $u$ is said to be a *median* of vertices $x, y$ and $z$ if $d_G(x,u)+d_G(u,y)=d_G(x,y)$, $d_G(y,u)+d_G(u,z)=d_G(y,z)$, and $d_G(x,u)+d_G(u,z)=d_G(x,z)$; equivalently, $u$ is a common point of suitably chosen shortest paths from $x$ to $y$, $y$ to $z$ and $x$ to $z$. $G$ is called a *median graph* if every triple of vertices has a unique median.

**Theorem** (Bandelt (1984)). A subgraph $H$ of a hypercube $K_2^k$ is a retract if and only if it is a median graph.

It is not obvious which of several possible generalizations of Bandelt's theorem will hold for products of complete graphs. This problem was solved recently by Chung et. al. (1987b) in the following way. Let $x$, $y$ and $z$ be vertices of a product $K_{n_1}\square \cdots \square K_{n_k}$ of complete graphs, with coordinate representations $x=(x_1, \ldots, x_k)$ etc. Define the *imprint* $\mathrm{imp}(x,y:z)$ of $x$ and $y$ on $z$ by setting the $i$th coordinate of $\mathrm{imp}(x,y:z)$ to be $x_i$ if $x_i=y_i$, and $z_i$ otherwise. Then:

**Theorem** (Chung, Graham & Saks (1987b)). An induced subgraph $H$ of a cartesian product of complete graphs is a retract if and only if $V(H)$ is closed under the imprint function.

Chung et. al. were motivated, in proving this theorem, by a graph game arising in the study of self-adjusting data structures.

## The Dynamic Location Problem

In most methods of storing data for use by a computer, some elements of that data will be more easily accessed than others. For example, if the data is in a list or tree, then the items at the top of the list, or near the root of the tree, may be more quickly retrievable than others. If data is located at physically distant places in a network, e.g. in different cities, then there may be a cost attached to accessing it which depends on the number of links between the processor and the data.

In recent years much attention has been devoted to the idea of revising a data structure according to how it is used, with the object of bringing the most-used items into the most cheaply accessed positions. Typically requests for data are buffered, so that the processor may know the next $k$ requests in advance; even though there is likely to be a cost to adjusting the data structure (say, by performing an adjacent transposition in a list, or by moving a file to an adjacent location in a network) it will be profitable to do so in certain circumstances.

The following game is intended to be a simplified (but still challenging to analyze!) model of this situation. A connected graph $G$ is fixed, with Player I (the "processor") located at a given initial vertex $v_0$. Player II begins by specifying a list of $k$ request vertices $q_1, \ldots, q_k$, where $k$ is fixed in advance. Player I may now move to a new vertex $v_1$, at a cost of $d(v_0,v_1)$¢, from which he must "field" the first request---at a cost of $d(v_1,q_1)$¢. Player II now supplies a new request vertex $q_{k+1}$, and Player I chooses a new position $v_2$ (at cost $d(v_1,v_2)$¢) from which to field $q_2$ (at cost $d(v_2,q_2)$¢). At all times Player I is provided with a "window" of $k$ requests, with which to make his next adjustment decision.

Player I's total cost after $N$ moves is

$$\sum_{i=1}^{N}(d(v_{i-1},v_i)+(d(v_i,q_i))$$

which he tries to minimize, while Player II tries to do the opposite.

There are many questions one can ask about this game---for example, what is the average value per move of the game for large $N$?---but the one asked by Chung et. al. (1987a,b) is the following: given $G$, what is the least value of $k$ such that Player I can play as well with a window of size $k$ as he could if he could see *all* the requests in advance. This value is known as the "windex" (presumably a contraction of "window index") of $G$ and denoted $w(G)$. If no such $k$ exists, we set $w(G) = \infty$.

For example, let $G = K_2$, with vertices labelled $a$ and $b$. It is not hard to show that with a window of size 2, Player I's correct strategy when located at vertex $a$ is to move to $b$ only if the next two requests are both for $b$. This works as well as any strategy, even when all the requests are known; on the other hand a window of size 1 is inadequate. Thus the windex of $K_2$ is 2, and similarly $w(K_n) = n$ for all $n>1$.

On the other hand, let $G$ be the complete graph on vertices $a,b,c,d$ with the edge $\{a,d\}$

removed. If Player I is sitting on vertex $b$ and sees $c,a,d,a,d,\ldots,a,d$ in the window, he must go to $c$ if the first unseen request is $c$ but stay at $b$ if that request is a $b$. Hence this graph has infinite windex.

It is not surprising, in view of previous comments, that the property of having windex $\leq k$ can easily be shown to persist under cartesian products and retracts. What is surprising, and not at all easy to prove, is that the complete graphs generate all graphs of windex $\leq k$ under these operations:

**Theorem** (Chung et. al. (1987b)). $G$ has windex $\leq k$ if and only if $G$ is a retract of a power of $K_k$.

The proof is too long to give here; it makes use of the metric representation theorem (above) of Graham and Winkler. In fact, Chung et. al. use the algorithmic part of the theorem as well. For, notice that $G$ cannot be a retract of a power of $K_k$ unless its canonical representation consists precisely of complete graphs on at most $k$ vertices (since the complete graphs are irreducible and nest isometrically). Having found such a canonical representation, one may check every triple $(x,y,z)$ of vertices (in polynomial time) to see that the imprint imp$(x,y:z)$ also lies in the image of $G$. If it does, and if $k$ is the largest order of the (complete) canonical factors (which, incidentally, is easily seen to be equal to the clique number of $G$), then by the above two theorems of Chung et. al., the windex of $G$ is exactly $k$. Hence:

**Corollary** (Chung et. al. (1987b)). The problem of determining the windex of a given graph is solvable in polynomial time.

Since to determine even the clique number of a given graph is NP-complete, we think it is impressive that there is a good algorithm for determining the windex---a parameter whose value even on $K_2$ is not obvious.

**References**

Bandelt, H.-J. (1984). Retracts of hypercubes, preprint.

Chung, F.R.K., Graham, R.L. & Saks, M.E. (1987a). Dynamic search in graphs, *Discrete Algorithms and Complexity* (H. Wilf, ed.), Academic Press, to appear.

Chung, F.R.K., Graham, R.L. & Saks, M.E. (1987b). A dynamic location problem for graphs, preprint.

Djoković, D.Z. (1973). Distance preserving subgraphs of hypercubes, *J. Combinatorial Theory (B)* **14**, 263-267.

Feigenbaum, J., Hershberger, J., & Schäffer, A.A. (1985). A polynomial time algorithm for finding the prime factors of cartesian-product graphs, *Discrete Appl. Math.* **12** #2, 123-138.

Firsov, V.V. (1965). On isometric embedding of a graph in a Boolean cube (in Russian), *Kibernetika* **1** #6, 95-96.

Garey, M.R. & Johnson, D.S. (1979). *Computers and Intractability: a Guide to the Theory of NP-completeness*, W.H. Freeman, San Francisco.

Graham, R.L. & Pollak, H.O. (1971). On the addressing problem for loop switching, *Bell System Tech. J.* **50**, 2495-2519.

Graham, R.L. & Pollak, H.O. (1972). On embedding graphs in squashed cubes, *Graph Theory and Applications* (Lecture Notes #303), Springer-Verlag, Berlin.

Graham, R.L. & Winkler, P.M. (1984). Isometric embeddings of graphs, *Proc. Nat. Acad. Sci. USA* **81**, 7259-7260.

Graham, R.L. & Winkler, P.M. (1985). On isometric embeddings of graphs, *Trans. Amer. Math. Soc.* **288** #2, 527-536.

Hell, P. (1976). Graph retractions, *Colloq. Intern. Teorie Combinatorie* **II**, Roma, 263-268.

Imrich, W. (1986). Embedding graphs into Cartesian products, preprint, Montanuniversität Leoben, Austria.

Nowakowski, R. & Rival, I. (1982). On a class of isometric subgraphs of a graph, *Combinatorica* **2**, 79-90.

Nowakowski, R. & Winkler, P. (1983). Vertex-to-vertex pursuit in a graph, *Discrete Math.* **43**, 235-239.

Sabidussi, G. (1960). Graph multiplication, *Math. Zeitschr.* **72**, 446-457.

Vizing, V.G. (1963). The Cartesian product of graphs, *Vychislitel'nye Sistemy* **9**, 30-43.

Walker, J.W. (1986). Strict refinement for graphs and digraphs, *J. Combinatorial Theory (B)*, to appear.

Welsh, D. (1982). Problems in computational complexity, in *Applications of Graph Theory* (R.J. Wilson, ed.), Shiva Publishing, pp. 75-85.

Winkler, P.M. (1983). Proof of the squashed cube conjecture, *Combinatorica* **3** #1, 135-139.

Winkler, P.M. (1984). Isometric embedding in products of complete graphs, *Discrete Applied Math.* **7**, 221-225.

Winkler, P.M. (1986). Factoring a graph in polynomial time, *European J. Combinatorics,* to appear.

Zaretskii, K.A. (1965). Factorization of an unconnected graph into a Cartesian product (English translation), *Kibernetika* **1** #2, p. 89.

## INDEX OF NAMES

This index includes entries for pages on which papers are cited even if the author is not explicitly named there.